Precious Objects

A Story of Diamonds, Family,
and a Way of Life

Alicia Oltuski

SCRIBNER

New York London Toronto Sydney

Scribner
A Division of Simon & Schuster, Inc.
1230 Avenue of the Americas
New York, NY 10020

Certain names have been changed.

First Scribner hardcover edition July 2011

Book design by Ellen R. Sasahara

Manufactured in the United States of America

1 3 5 7 9 10 8 6 4 2

Library of Congress Control Number: 2011005573

ISBN 978-1-4165-4512-5
ISBN 978-1-4391-7170-7 (ebook)

For the members of my family, who have lent me
their lives and given me mine, and most of all,
for Uri

Contents

Precious Objects

Chapter I

Our Brightest Blazes

My father handed me the chest pack. It was a tan pouch the size of a small outstretched hand. I took off my jacket and slipped the pack over my tank top, on my right shoulder, where the pack rested between my breast and armpit. I slid on my jacket and zipped up. My father checked me for bulges.

Inside the pack, I had carefully placed the gold bracelet, a thick Victorian piece decorated with flowers, whose petals were made of sapphires and whose centers, of diamonds. A few of the diamonds were missing. I was to pick up the replacements at a jewelry exchange down the street and carry everything to the setter, who would fit the gems into the bracelet.

I was twenty-two, and this was my summer job. My father paid me what I had made teaching English as a Second Language at college—an amount that, two months earlier, he had dismissed as underpayment. "If they went anywhere else, they would be paying at least twice as much," he'd complained.

Now, with the pack secured, I stood next to his safe, a massive silver-gray box that hovered like a crouching giant at forty-two inches high. In addition to two doors and a combination code, there was a key formed from a long rod with an encoded tip that had to be reassembled with each use.

Inside the safe lay dozens of black boxes with windowed tops through which the diamonds sparkled. There were black trays filled with gleaming old jewels: a musky cameo brooch, a diamond pin with a hunk of aquamarine, a velvet necklace with a green peridot gem pared into the shape of Medusa's head, and a flower diamond brooch that shivered when shaken. Once, I even spotted a gold chain with dangling angry-looking warriors' faces painted onto blocks of ivory.

Most of my father's jewelry is old; it suggests a time in which not only taste but also bodies were different. He owns wristwatches so slender they look as if they'd break if one ever checked the time, and earrings so heavy I could imagine them tugging on the lobes of giants.

On top of the safe is a framed picture of his brother, Steve, sitting tranquilly on a boat in the middle of a lake. Near the desk where my mother works when she comes in to assist my father stand six photographs depicting my parents on a night out in Germany, where they grew up and lived before I was born, and a shot of our family at my sister's bat mitzvah, my mother draped in diamonds. Behind my father's table is an old photo in which his business friend Lester sits on his lap while my father wears an exaggerated expression of pain.

Though my father's appearance is orderly—he believes in tidiness, in general, and clean fingernails, in particular—his desk is often cluttered with papers, stones in boxes, and trays of jewelry. He sits the farthest possible distance from the door, so that he can keep watch over who enters his office without being within their immediate reach. Next to him is a row of windows through which the masses of foot traffic that flood New York's diamond district are visible, or would be if he didn't always keep his shades drawn. Cameras displaying the hallway outside hang from the wall, and beneath his desk, always at his fingertips, is a red panic button that summons the police.

The office itself is like a safe. To leave that day, I passed through three doors. First, a standard gray lightweight; then a bulky metallic one, containing a bulletproof window with a bank slide for quick and secure exchanges; and finally, the outermost door, wooden and professional-looking, bearing the company name, Oltuski Brothers, though there is only one brother in the business. The wooden door and the metallic one are set up so that when one is opened, the other locks automatically, to slow getaways in the event of a robbery.

Having two or three office doors isn't unusual in the diamond district, but this labyrinth of security particularly suits my father. This is a man who made his entire family wear surgical masks for weeks after my sister's birth. He wouldn't let me near the playground sprinklers because he'd heard somewhere that they could spread hepatitis. He forbade me from marching in the Israeli Day Parade with my classmates and all the other Jewish school students in the tristate area. "Too many Jews in one place," he said. A perfect target for terrorism. A perfect target was why I couldn't go on my high school's graduation trip to Israel. The fortification of the diamond business seeped into our private lives, and my father hid us from the world just as fiercely as he hid his stones.

Because he wanted to keep me safe, he didn't dress me in

his diamonds. In fact, I wasn't allowed to pierce my ears until I was sixteen, because he'd heard somewhere that muggers had torn a woman's earrings straight out of her lobes on the streets of New York. Any gem I possessed was kept in my father's safe along with my baby teeth. I never minded any of this until once, in synagogue, his friend looked at my arm and said to him, "What? Not even a little bracelet?" and I burned with shame.

As I left the office, I passed by the framed decorative cards my mother had hung in the corridor to make the white, fluorescent-lit room less sterile: a cat wearing a diamond tiara; loose diamonds glimmering inside an egg carton; a cartoon diamond that says *You Rock*; and a Samuel Johnson quote: "Our brightest blazes are commonly kindled by unexpected sparks!" She believes that if the office looks nice, people will be inspired to buy.

In the lobby of our building I pushed past the turnstiles by the security desk, through two more doors, and then I was on the street with five thousand dollars' worth of jewels strapped to my body. I was nervous. Four years as an English major hadn't prepared me for diamond deliveries. And neither had my father. He applied silence to as many parts of our life as he could, especially diamonds.

He was always hiding things. If a stranger asked what he did for a living, he never said he was a jeweler. Instead, he made up a different profession. Sometimes he sold insurance. Other times he was a product representative. But I didn't have an alternate persona to slip into. Instead, I crossed Fifth Avenue from the quieter east side, where my father's office building resides, to the west, and tried to act casual. But when you are carrying diamonds, each block is a continent.

I was greeted by two colossal diamond-shaped lamps

built atop giant metallic stanchions. These columns guard all four corners of Forty-seventh Street between Fifth and Sixth Avenues and signal the borders of the district's most important quarter. Crossing was like going from an American suburb to a medieval marketplace. Hired men and women with signs over their chests and flyers in their hands walked the street calling, "We buy gold," "We buy diamonds." Dealers convened in pairs, making verbal contracts, gossiping about who bought the three carater for fifty thousand, or how much that special art deco necklace sold for. Armored trucks stood parked on the street while couriers sorted diamond deliveries under the scanning eyes of guards with guns, and the civilians of Forty-seventh Street, the customers, pressed their noses against the ground-floor window displays.

Inside the windows, earrings hung like fruit from little cushion trees. Necklaces were draped around plastic busts. Gold chains dangled from hooks on suction cups. Diamonds were peppered onto brooches and bracelets like garnishes. Glittering hearts, skulls, starfish—whatever you desire.

Deeper into the street, I reached the exchange at 10 West Forty-seventh Street. It felt like a casino. Men and women stood around waiting for their lucky break. Dealers called out numbers—the prices and dimensions of stones. The room was packed so densely, they hardly needed to move to show someone a diamond. I squeezed through the narrow aisles lined by jewel-filled booths. Two men played cards over a showcase. Another sucked on an unlit cigar, and a few uniformed guards milled around aimlessly.

Each company in the exchange had a display case and a booth the size of a half bathroom. Dealers sat on stools or folding chairs, or leaned onto their cases. On the cubicle walls hung dollar bills, family pictures, and signs explaining refund policies. Some booths, even those of non-Hasidim, had a picture of the famed white-bearded Lubavitcher rebbe, whom many devotees, undeterred by his death in 1994, believe to be the Messiah. Sometimes I wondered if the pictures were meant to draw business from his followers.

The exchange was lit by long white lamps that hung diagonally from the ceiling, and track lights that hovered above the booths, but despite all these efforts at illumination, the room remained somewhat dim. Its floors were dusky, covered in black and yellow shapes that resembled small rocks. The hum of steady conversation filled the hall, frequently punctuated by telephone rings and, every once in a while, the sound of heated bargaining or a steam machine blasting dirt off a jewel.

Toward the front of the exchange, in a corner booth, I spotted a black arced plastic sign that simply said *Ginsburg,* the name of the man who had borrowed my father's stones. His booth was sprawling compared to some of the others. Its three display cases were filled with jewelry in so many different colors and styles—painted brooches, blue and green gemstone rings, butterfly shapes and gilded leaves—that it resembled a high-end carnival prize stand. Behind the showcases was a chain of space-efficient desks covered in paperwork, and above them, gray cabinets. A glaring diamond light was mounted to the wall. A handful of employees, including Ginsburg's wife, negotiated the narrow space. I stood on the outside of the booth, in the aisle.

Ginsburg (I promised him I would use a pseudonym) sported an elegant mustache and goatee. His eyes were dark and pointed, like periods. He didn't have all of his hair, but a large part of his head was covered with his black *kippah.*

I told Ginsburg I was from Oltuski.

"You're *from* Oltuski or you *are* Oltuski?"

It isn't hard to see that my father and I are related. I inherited a variation of his downturned eyes, which can make us look sad. We are both on the short side, and have ears that salute at an angle, rather than falling neatly into line with our temples. Ginsburg said to tell my father he was sorry that he had not been able to sell the stones.

At any given moment on Forty-seventh Street a dealer may be in possession of hundreds of thousands of dollars' worth of another man's diamonds. Usually, he has not paid the owner any collateral, only his word. He receives the diamonds in an envelope with a name on it. Sometimes the owner doesn't even ask for a receipt to prove that the merchandise is his, although my father, not surprisingly, had. In this business, everything works on credit, loan, and trust. Had Ginsburg been successful in selling my father's diamonds, both would have made money. Instead, my father would use the diamonds in the Victorian bracelet I was carrying.

I gave Ginsburg the receipt. He handed me the plastic baggie of stones. I tucked them into the pouch as quickly as possible and, afraid to draw attention to myself, resisted looking around to see if anyone was watching me. When I had secured the merchandise as inconspicuously as I could, I left the exchange and made my way down Forty-seventh Street to the setter's office, on the eighth floor of a small, nondescript building. There was no suite number or business name on the outside of the office, so my father had described it to me by its damaged doorbell, which dangled from the post.

The setter's workplace was the size of a modest kitchen and smelled of chemicals. A poster of Pope John Paul II hung on the wall above his desk. It was the only place on

Forty-seventh Street I'd ever seen a religious item that was not Jewish. Then again, it was one of the few times I'd met a colleague of my father's who was not Jewish.

The setter's floor shone slightly from layers of gold dust that had fluttered from his table over many years of work. Jars of cream and gel, and a multiplicity of pliers, adorned his desk like an alchemist's laboratory. Some of his potions seemed sorcerous, like the protective jelly he slathered on a ring before molding it. After glossing the ring with a paintbrush, he would stick it into a flame sprouting from the burner on his desk, and the flame would turn an unnatural shade of fluorescent green. Beneath his table were foot pedals that operated the various drills he used to shave down his customers' precious metals. The drills' cries sounded like animals shrieking underwater.

As though to offset his noisy tools, the setter himself spoke in a mellow voice, tinged with a Polish accent. He was a tall, skinny man with slender legs, but his most striking features were the knuckles of his thumbs and index fingers, which protruded orblike, as if marbles had been placed beneath his skin. Over decades of jewelry setting, his hands had evolved for efficiency. His thumbs were spaced unusually far from his other fingers, the skin between them stretched thin as a web. The tendons on the undersides of his forearms braced themselves so markedly, they resembled the prongs of a fork.

The setter didn't have much to say as he accepted the plastic baggie and bracelet. I repeated the instructions my father had given me, pointing to the holes in the bracelet's flowers where the diamonds belonged, just in case. The setter indulged me, smiling politely while he said yes, but, of course, he already knew all of this.

Other than the floor, his office didn't sparkle or shine,

though he must have been in possession of dozens of jewels. Most likely none of them belonged to him. Setters don't own the pieces they mend. Like foster parents, they receive the gems, hold on to them for a while, tend to them, and then let them go. Though they hold vast riches in their hands, they themselves can be poor. The competitive ones keep their commissions down to only two percent. Sometimes they're nicknamed "diamond-studded paupers." A safe full of jewels does not necessarily make a wealthy man.

When the setter finished the job, my father would pay him and try to sell the bracelet. No one would be able to tell from looking that it was this setter who had worked on the bracelet, just as no one would know that it was my father who chose the diamonds for it. After a stone is set into jewelry, it is almost impossible to decipher who harvested the gem from the river shore, who carried it from the wilderness to the market, who cut it, who polished it, and who made the match between the stone and its setting. Once a diamond becomes a piece of jewelry, the fingerprints of all these people disappear.

My errand done, I returned to the quiet of my father's office and worked on organizing the records of his retirement accounts. The papers sat next to his diamond scale, under colorful pictures of gemstones my mother had put up, and dated back to the early nineties, when my father's daily pursuits were still secrets to me.

Even then, I'd known that I wouldn't go into the business. He had never encouraged me to. He wanted a different life for me, one in which my next paycheck would not be dependent on the whims of some dealer or rich woman. But here I was, working in his office, because I needed to understand the street, the stones, and the people that occupied his life.

That evening, as on every other, my father put his jewels into his safe, punched the code to his alarm, and turned off the lights in the vestibule between his outermost doors, standing for a moment in the darkness. Before locking up manually, he kissed the mezuzah, that small rectangular box on the doorpost containing the holiest of Jewish prayers—one of the many rituals he performed daily to protect himself, his possessions, and the people he loved. Then he started walking very slowly down the hall, as though inching away from a crime scene, so that he could simultaneously make progress toward the elevator and hear the sirens from behind his door that indicated his office had been secured. In the narrow corridor, his eyes automatically scanned the two circular mirrors mounted on the wall that showed if anyone was lingering around the bend. By the time he left, he was almost always the last one on his floor.

In the lobby, he bade the guard on night shift good-bye. Then he stepped outside into Manhattan's heavy August air, and left behind the fluorescent glow of the diamond columns, the storefront signs coaxing pedestrians in with *We Buy Gold and Diamonds,* the unmade deals, and all the jewels.

Chapter 2

Glitter Becomes a Currency

It was only after I stopped working for my father that I started thinking about the origins of the diamond business. The name De Beers had been ringing in my ears like a beloved ancestor's since I was young, and I'd even heard the ominous term "the Syndicate" being tossed around. I grew up listening to the occasional diamond talk around the dinner table or over the phone, and if I really rummaged through the files of my mind, I probably would have intuited that the bulk of the story had begun somewhere in Africa. But I didn't know the story itself.

On Forty-seventh Street, dealers talked about the price of diamonds, whether the owner was willing to come down, and if a stone's make was pretty enough to pay a premium—not how diamonds first came to be considered valuable or romantic, not how diamonds first came to be.

On an unforgivably hot day in 1867, a teenager named Erasmus Jacobs found a shiny stone near the Orange River of

South Africa. He brought it home and gave it to his sister. Pebbles were good for a game called "five stones." The rock made its way into the Jacobs family collection, where a neighbor of theirs spotted it. He seemed to like it, so Erasmus's mother gave him the stone.

The neighbor chafed the rock against a window to see just how much he liked it—in other words, whether it was a diamond. But he couldn't be sure, so he sold it to a traveling merchant, who took it to a commissioner, who dispatched it to a mineralogist, who declared the stone a diamond.

At the time, Brazil was the world's foremost diamond producer, a title it had taken from India in the mid-1700s. But this was the beginning of South Africa's reign. Erasmus Jacobs's stone eventually weighed in at twenty-one and a quarter carats. It ended up in the hands of the Cape governor, who showed it off at the Paris Exhibition.

Two years later, another boy found a diamond. This one weighed eighty-three and a half carats and would come to be called the Star of South Africa.

"Gentlemen, this is the rock upon which the future of South Africa will be built," said the colonial secretary when the gem made its appearance at the Cape Assembly.

All of a sudden, South Africa was a diamond destination. First came the diggers. Men who were not diggers became diggers. Men who were not diamond prospectors became diamond prospectors. They came from North America, Europe, Australia, even the ocean; seamen left the waters behind for a more promising career on African land. In the beginning, they stayed close to the Orange and Vaal Rivers, since diamonds were thought to exist only near water. They set up camps, which flowered into small digging societies with newspapers, law enforcement, and even a hotel.

The move to dry land started with just a few, but soon others began to realize that it wasn't only the rivers; South Africa was bursting with the glimmering stones. Even the earthen walls of farmers' houses shone with diamonds. So the diggers left the riverbanks for the inlands. On the farm of a family named De Beers, they hit the jackpot. By 1871, the orderly farm had turned to chaos. Along with hundreds of diggers, the diamond rush brought with it inebriates and criminals. The De Beerses were not fond of what diamonds had done to their land, so that year they sold their property and left. The family earned only six thousand guineas, but their estate would turn up millions of dollars' worth of diamonds over the course of the next hundred years. The mine that originated on their ground came to be called the De Beers Mine.

Very soon, other farms in the area were ravaged. There was Bulfontein Mine and Dutoitspan, but greatest of all was the Kimberley Mine, a site about a mile from De Beers that was christened the Big Hole.

Over time, the owners of digging claims dug so deep that the earth began to change color and character—from yellow to blue and from docile to tough—so deep that the barriers between one man's plot and another's began to fall away. The laws that governed diamond digging changed, too. By 1876, the maximum number of plots one person could own had risen from two to ten. This set the stage for another kind of rush. This time it wasn't just about collecting diamonds. It was about acquiring as much diamondiferous land as possible. This development was necessary to the future of the industry: if hundreds of diggers merely sold their finds individually, providing discounts to compete with one another and unloading as many stones as possible, the market would be overrun, and diamonds would eventually be as cheap as gravel.

Two men surfaced as the principal contestants in the race of mergers that would shape the trade. One of them was Cecil Rhodes, who lent his name to the Rhodes Scholarships, established by the University of Oxford according to Rhodes's will, and for a time, two countries, Rhodesia and Northern Rhodesia (now Zimbabwe and Zambia). He remains notorious for his fanatical and racist imperialism, but when he first arrived in South Africa at the age of eighteen, he had neither fame nor notoriety. He suffered from asthma and a weak constitution. At nineteen, he had a heart attack.

Rhodes had come to South Africa at the invitation of his older brother to help out on a cotton plantation, but it wasn't long before he made his way to the mines and staked out claims at De Beers that proved fertile. About three years into his South African adventure, Rhodes returned to England to attend Oxford. There, he was exposed to the imperialist John Ruskin, who, without a doubt, stimulated his supremacist ideas about the British nation, a nation that, in Rhodes's opinion, would improve the globe by occupying as much of it as possible. As for Africa, he wrote, "it is our duty to take it."

While at Oxford, Rhodes traveled back to South Africa frequently. By 1880, he and his financial backers owned a considerable hunk of the De Beers Mine. That year, he established the De Beers Company. He also plunged into politics, joining the Cape Colony Parliament in 1881.

In this new land of diamonds, politics and gems overlapped. Rhodes was not the only one to try his hand at governing; his future archrival, Barney Barnato, was elected town councillor of Kimberley. Barnato was a Jew from London's inner city who was known to recite Hamlet's "To be or not to be" soliloquy in a handstand. When Rhodes first set foot in Africa, Barney was still performing music hall routines with his brother Harry in their hometown. Their real last name was Isaacs, but a cute show house line that Harry used ("and Barney too!") eventually evolved into Barnato.

After brief stints as a boxer, comedian, stationery merchant, and cigar seller, Barnato gave diamonds a go. He started out by paying diggers to let him scour their exhausted land on the off chance of finding a neglected diamond. Eventually, Barney and his brother bought a piece of the Big Hole. They made their purchase of four Kimberley plots at exactly the right time; it was after 1876, so the digging maximums had already changed. Moreover, the ground was almost all blue, a color most diggers feared was a sign of depleted land. Not Barney Barnato. He put his trust in a hypothesis that diamonds emerged from volcanoes deep in the earth, closer to this new blue ground than the preceding yellow. He was right.

Every once in a while, history is irrevocably changed by the chance encounter of scientific discovery and financial ingenuity. As more and more stones surfaced, geologists realized they were coming from much deeper in the earth than river streams. When the plots were burrowed deep enough, they presented large amounts of diamonds. A few of these plots belonged to the Barnatos, and sure enough, stones of up to twenty-five carats appeared in their four Kimberley claims. This new blue ground, it turned out, was crystallized magma that had carried diamonds upward from the depths of the earth. Barnato had hit upon the source—blue was beautiful.

He used his profits to buy more digging ground and started the Barnato Mining Company, which became one of the top four controllers of the Kimberley Mine. Next, he set his eyes on the other three and bought their shares. He was able to take over two of the companies, making him chief shareholder of Kimberley Central, the firm that now dominated the Big Hole. Kimberley was basically Barney's, which officially put Barney in Cecil Rhodes's way.

As his first order of business, Rhodes tried to buy the one major Kimberley company that Barnato had not managed to procure: the French Company. He solicited funds from London investors, but shortly after he put in his offer of £1.4 million to the French, Barnato countered with £1.75 million.

Rhodes knew that he and Barnato—and their financiers—could go on and on, throwing millions of pounds at the French Company holders. So instead of merely offering a higher figure, he approached his opponent with a deal: all of the French Company would be Barnato's, in exchange for allowing Rhodes to act as middleman. Rhodes would buy the French Company and resell it to Barnato for £300,000 and one-fifth of the shares in Kimberley Central. This, Rhodes told Barnato, would spare him the embarrassment of having lost the game. Barney accepted, figuring he was gaining the entire French Company and holding on to the majority of the Central. But Rhodes was scheming. After the deal went through, he quickly acquired any Central shares that were not in Barnato's hands, including those of a number of shareholders Barnato thought were on his side. He bought his way into sixty percent of Kimberley Central.

But Rhodes was after something bigger than business. He knew that minerals were the key to a kingdom's fortune. He wanted the earth to come under the dominion of the British Empire, and he wanted stones to help pay for the venture. This wasn't just two men having it out, this was a tycoon and a conqueror going to war.

The battle finally ended at four o'clock one March morning in 1888 during a face-to-face meeting. After hours of persuasion, Barnato finally agreed to let Rhodes purchase his stock. That same month, Rhodes founded De Beers Consolidated Mines. He and Barnato were challenged in the Supreme Court of Cape Colony by a cohort of Barnato's former shareholders who were not eager to be incorporated into De Beers. The loophole that Barnato and Rhodes used was to liquidate Central and to sell Barney's overwhelming share to De Beers as individual parts. In 1889, they issued the most valuable check the world had ever seen: £5,338,650. For the rest of his days, Barnato was a director of De Beers, but it was Rhodes who would determine its future. After the merger, Rhodes decreased excavation in De Beers Consolidated Mines. If diamonds were not scientifically rare, De Beers would make them seem that way.

The company went on to adopt the strategy of selling rough diamonds (diamonds that have not been cut and polished) to a handpicked group of sellers in London, referred to as the Syndicate. De Beers learned not only to control supply and demand, but to divide supply into two fronts: how many diamonds came out of the earth, and how many diamonds were released into the market.

During Cecil Rhodes's reign, the company controlled ninety percent of the world's rough diamonds. But the new De Beers was not a victory for everyone. Rhodes dramatically shrank employment. According to Tom Zoellner, author of *The Heartless Stone,* mine regulations helped lay the groundwork for apartheid South Africa. In the city of Kimberley, a new law compelled servants (meaning mostly black people) to keep work papers with them at all times and show them to any white person who asked to see them. Upon exit from the mines each day, black workers had to take off all of their clothing, and guards searched their bodies for stolen gems. Soon, rather than being allowed to leave at all, black miners were packed into compounds, where they remained for the entirety of their employment. These rules weren't entirely new. There are sources that show abuses like naked searches even before the consolidation of De Beers. In the early 1880s, despite protesting, white workers had to undergo searches, as well.

Rhodes never intended the diamond company as the end of his conquests. The man hungered for more of Africa and set his sights on Matabeleland, which was known to harbor gold. Soon after the De Beers consolidation, Rhodes got a charter from London for the British South Africa Company he'd created, which allowed him vast expansionist rights on the African continent. The next year, he became prime minister of Cape Colony. His delegate would later trick the Matabele king Lobengula into essentially signing over his territory. The land was soon flooded with whites in search of precious metal, and eventually it was referred to as part of "Rhodesia."

Barney Barnato didn't fare as well. By late 1896, several personal and financial misfortunes were taking their toll on him. He was drinking. He was looking for diamonds in the wall of his home. He was begging to be let into a neighbor's house because he was convinced someone was after him.

Barnato achieved his last bit of fame upon his departure from South Africa. One afternoon, on a steamship headed from Cape Town to England, he threw himself into the sea. Some of the obituaries referred to him as the Diamond King, but although he remains an integral part of De Beers history, the company continued with a life of its own long after Barney's was over.

While De Beers' early days are defined by Barnato and Rhodes, most of its existence has been spent under the leadership of the Oppenheimer family. In 1929, Ernest (later Sir Ernest) Oppenheimer became chairman and began a diamond dynasty. His son, Harry, succeeded him.

If Cecil Rhodes introduced the art of controlling diamond supply to De Beers, then Harry Oppenheimer perfected De Beers' manipulation of consumer demand. In the late thirties, right before the war, Harry and his advertising agency, N. W. Ayer, began to craft one of the most cunning feats of marketing history—a campaign that would equate diamonds with love and eternity.

Before Harry teamed up with Ayer, American engagements were responsible for the majority of the company's diamond sales, but the gems in these rings cost a measly eighty dollars, on average. Harry aimed to change that.

Ayer used movies, newspapers, and radio to plant diamonds into the daydreams of Americans, casting the stones as emblems of romance and glamour. On-screen, beautiful women received the gemstones. Newspapers printed pictures, fed to them by Ayer, of the upper crust sporting diamonds on their hands and necks. Magazines ran colorful advertisements, made up of a visual—such as a painting by Dalí or Picasso—sometimes a literary quote, and often, a written scene of love, containing a lesson: those first moments of romance are brief, but diamonds last an eternity. A young couple in love are wise to preserve their passion in a stone.

One such ad read:

> Endless they seem to young people caught up in love's first fine awareness and yet already they are fleeing. . . . Fortunate indeed are those who carry with them out of the flare of their dreams some tangible object to grace a lifetime's mature happiness with the memory of this first sweet halcyon. Most frequently, such a talisman lies in their engagement diamond. No other treasure of earth or sea which they may acquire in later life will ever have one-half such precious significance for them.

At the bottom was a little guide showing preferred sizes for unmounted ring stones.

Perhaps the cleverest of these advertisements addressed the ongoing Second World War head-on. "Industrial diamonds—a key priority for high-speed war production—come from the same mines as gem stones," they informed readers. "Millions of carats are used in United States industries today. The occasional gem diamonds found among them help defray production costs for all these fierce little 'fighting' diamonds. Thus, there are no restrictions on the sale of diamond gems."

In some advertisements, young women waited for their men off at war, with, thank goodness, diamonds on their hands to remember them by. Although these ads spoke of the men's return and referred to the ring as a symbol of the couple's love, everyone knew that not all fiancés and husbands returned home. The underlying message was powerful: humans are mortal, diamonds are not.

The ads always said *De Beers* at the bottom, but they didn't need to, since the company—which engaged in mining, not retail—had no real competition. De Beers provided almost all retail diamond sellers with unpolished stones. This was a campaign of ideas, the idea that romance came in different cuts, colors, and varieties of diamonds. And the idea caught fire. In three years, between 1938 and 1941, American diamond sales went up fifty-five percent.

It was not just the men and women of marriageable age who were being won over. De Beers toured the States, holding assemblies at high schools about the history of diamonds, farming an entire generation of future consumers.

Then, in 1948, just three years after the world had almost come to an end, De Beers announced that *A Diamond Is Forever.* And so it was.

Nineteen forty-eight was also the year that South Africa officially became an apartheid nation. De Beers' relationship with the country, and the African continent in general, is complex and multifaceted. The Oppenheimers themselves have always been progressive citizens. The election of 1948 won Harry a seat in Parliament as a representative of the United (antiapartheid) Party. Harry condemned apartheid, and when the question of black enfranchisement polarized the party, he banded with the liberals, who didn't want to rule out the possibility of black voting rights. Eventually, he left the United Party and heavily subsidized the new Progressive Party.

Nevertheless, by the 1960s, apartheid proved glaringly perilous to the future of De Beers, for this was the age of African nationalism and independence. Between 1957 and 1964, the Gold Coast, Tanganyika and Zanzibar, and Sierra

Leone all won self-rule. De Beers' strategy at the time was to wolf down as much of international diamond production as it could, which meant that the company depended upon its relationship with the outside world, a world that South Africa was enraging. The diplomatic rapport between South Africa and the new African nations was nonexistent. Then, in 1963, the Soviet Union ordered an embargo of South Africa, which most of the African continent rallied behind. This was a huge roadblock to De Beers. Several African countries were home to valuable diamonds the company needed to buy, and the Soviets had discovered a large diamond reserve in Siberia that could jeopardize both the price of diamonds worldwide and De Beers' place at the top of the pyramid.

Harry Oppenheimer understood that it would be impossible for countries such as Ghana and Sierra Leone to openly sell to De Beers. But they could sell to companies in Europe, which could then inconspicuously direct the diamonds on to De Beers. So Oppenheimer conveniently allowed "independent" companies to pop up. The diamonds could then be sold to firms located in more palatable countries, such as Luxembourg. Leaders of the gem-producing countries in Africa were fully aware of the cover-up, but it was too good an opportunity to pass up, especially for a developing nation in need of the income.

Oppenheimer worked out a similar clandestine understanding with the Soviets, who would sell every last Russian diamond to De Beers, ensuring that De Beers remained in control of global diamond prices. And, as the reporter Edward Epstein points out in his book *The Diamond Invention,* the diamond-selling countries could publicly attack South Africa while secretly doing business with its largest company.

Over the years, in their London vaults, De Beers stockpiled diamonds said to be worth billions of dollars.

Ten times a year, the Diamond Trading Company (DTC), the company within De Beers that distributes rough stones, sells its diamonds to customers at offices in London, South Africa, Botswana, and Namibia. The transaction is called a "sight," and the customers are "sightholders." Not just anyone can be a sightholder; diamond firms can apply to buy a certain class of rough, but De Beers decides whom they want as sightholders. The company chooses its sightholders based on various benchmarks, including how well the applicant will sell its diamonds. All sightholders must adhere to the legal and ethical standards set forth in the company's Best Practice Principles.

Ten times a year sightholders travel from all over the world to De Beers offices to pick up their rough diamond assortments in plastic boxes that say *DTC* on them. These ten sights mark the beginning of a process that enlists innumerable members of the industry. For decades, these rough stones trickled down from sightholder to cutter or dealer (sometimes on Forty-seventh Street), from dealer to broker, broker to other dealers, dealers to small or large shops, and finally from shops to customers. Sometimes the dealer skipped the broker and went straight to another dealer, but more or less, this was the diamond pipeline, year in, year out.

Then, in the 1990s, the pipeline began to constrict. De Beers decided to renovate. It had been facing competition from two fronts. In 1996, it got into a row with Australia's Argyle Mine, the majority of whose goods it had been buying since the eighties. The quarrel led Argyle to end its selling agreement with De Beers, instead allowing Indian cutting houses to buy its diamonds. De Beers, in turn, released a host of stones similar to the ones Argyle sold in India, presumably to warn the mine by devaluing its product. The plan flopped. The gems Argyle had been producing were brown—typically not a highly desirable color—but Argyle had been successful in creating a marketing campaign. So successful that the appeal of their diamonds, which were no longer brown but rather "cognac" or "champagne," couldn't be destroyed by De Beers' surfeit trick.

Canada was De Beers' second front. Beneath its icy terrain, diamond mines were being unearthed by other companies. Among them was the major BHP Minerals, which wasn't about to sell out to De Beers. De Beers was still considered a monopoly in America, where BHP owned coalfields. An alliance with the diamond giant could endanger BHP's business in the United States. De Beers' old strategy of buying out alternate diamond sources was going to be very difficult.

The late nineties brought a change of command. The De Beers mantle passed from Harry Oppenheimer to his son, Nicky. In 1998, a financial consulting firm gave De Beers a radical suggestion: they would be better off without their trademark policy of stowing away diamonds. A stockpile meant stable prices for everyone, including competitors like Argyle. By 2001, the company had sold a quarter of the contents of its vaults.

Next, De Beers dipped into a market it had never before touched: diamond retail. In addition to supplying rough to sightholders, it started offering polished, ready-to-wear De Beers–brand diamonds to the consumer in 2005. The company even came up with a special "Forevermark" brand, under which it markets high-caliber diamonds that have been mined using stringent ethical and environmental practices. As of 2010, Forevermark diamonds are cut only by "master craftsmen" and sold exclusively in Asia. The gems are engraved with a "Forevermark," a tiny square symbol, abstractly reminiscent of a diamond, and below it, an ID number, discernible only with a machine that only Forevermark jewelers carry. Branding has an important implication beyond the mark of luxury: as more companies carve names and numbers into their stones, diamonds will no longer be the anonymous, untraceable objects they once were. This trend holds enormous significance for consumers who want to be certain they are not purchasing jewelry from ethically questionable sources.

But Forevermark is scary to some of the people who share the industry with them. For decades, the advertisements De Beers printed in magazines and aired on television had been meant to get people to buy any and all diamonds. But now that De Beers is also a brand, it is employing its name for its own merchandise, a name that has been whispered into the ears of consumers for decades. The retail segment of its business involves no middlemen: no sightholders, no wholesalers, only De Beers and its customers. The 2005 launch store opened on Fifth Avenue, eight blocks north of Forty-seventh Street.

I needed to know what this was doing to the district, so I went to David Abraham, the former vice president of the Diamond Dealers Club and a dealer on Forty-seventh Street. David grows a five o'clock shadow before three p.m., wears big round glasses, and is the holder of so much information about the industry—and a hyperenergy that cannot apparently be satisfied by a life in diamonds—that he commonly interrupts himself multiple times a sentence. According to David, "people were very concerned" at first, but "once the store opened and the trade noticed how little of an impact it really had, it became a nonissue."

But I've also heard dealers grumble about De Beers cutting out the middleman. Its retail stores mean that there are fewer opportunities for them to participate in the jewelry chain. "De Beers, they like to take the profit all for themselves," said one gem and diamond merchant I spoke to.

In the process of revamping, De Beers became a private company, a status that grants it greater independence in decision making. It also cut down on sightholders from more than one hundred to about eighty. The stipulations for becoming and remaining a sightholder became stricter. If you were part of this elite group, you were encouraged by De Beers to spend ten percent of your revenue on advertising, up from an approximate one percent. All this was part of a move toward marketing the diamond within the broader category of luxury items. It wanted people to think of De Beers diamonds in the same breath as fancy cars and designer perfume. During the makeover, it partnered up with LVMH, the group responsible for promoting Louis Vuitton. It also urged sightholders to imprint their diamonds with the De Beers logo. It is no longer good enough to sell just a diamond.

But none of this changes the fact that De Beers set the historical stage for an entire industry—from sightholders in London to polishers in India to small dealers like my father in New York. All of them personalities. That's what I learned: that diamonds have always been a business of big personalities, their clashes and ambitions, and their innovations. I saw them all the time in the district where my father traded. Jokesters, instigators, and artists. I just hadn't known that they'd been there since the beginning.

Most likely, some of my father's antique diamonds are old De Beers. It is even possible that his were mined in the age when the earth began cashing out, when South Africa was repopulated by men seeking wealth at the edge of waters, when Rhodes and Barnato had it out, which all happened because a young boy near a riverfront let something beautiful catch his eye. After that, it wasn't long before glitter became a currency.

Chapter 3

The Deal

Summer's heat has already abandoned the city when my father and I walk from his office to the exchange in September of 2008. It's been about two years since I worked for him. Today, he is wearing the chest pack, filled with jewelry in small plastic baggies.

An armored truck stands near 10 West. The company has a small strip of the street cordoned off. Deliverymen hang around the curb with guns. My father is on his way to look at a fifteen-carat diamond. He and his friend Roy, the owner of a diamond-cutting and -selling operation, are thinking of partnering up to buy the stone.

Roy is a tall man with a gruff voice who bears a resemblance to Walter Matthau, only with paler hair and freckles. He and my father talk on the phone multiple times a day, discussing how much to buy or sell gems for. Sometimes they stop by each other's offices with a diamond, in need of a second pair of eyes.

"Hey, Grumpy," said my father when Roy picked up the phone. Instead of just listing the stone's measurements and quality, he asked Roy to think back to a different one they'd purchased together in the past. "You know that stone we bought and sold—the eleven oh five, you still remember it? Well, try to remember it. . . ." He reminded Roy of the diamond's pigmentation and described its flaws: "It was an O color with a white VS-two on the side. . . . We bought it cheap. . . . If you saw a stone like that today again, what would you pay? . . . Would you pay more? . . . Yeah? How much more?"

Soon, my father started adjusting the specs. "What if it's a fifteen carat. . . . Okay, a fifteen carat with a black VS, black VS-one maybe, maybe two. . . . But it's a, you know, smaller, it's a VS-one. . . . Come on, it's a fifteen carat." My father began to get annoyed. A little argument broke out. "Do we wanna give the same thing for a fifteen?" He didn't think they could get a fifteen carat at the same price they'd paid for an eleven, even if its clarity was somewhat inferior. The diamond in question was a large yellow Cape stone, the color of light beer, but it had a small black flaw—*very slightly included,* VS for short. Finally, my father revealed to Roy who had offered him the stone. It was a dealer at 10 West, the exchange.

On our way to see the diamond, my father and I take a detour down to Twenty-fifth Street to preview an auction at the Tepper Galleries, the oldest auction house in New York. We pass its scopic windows fronting Twenty-fifth Street, just around the corner from Park Avenue. Large black letters reading TEPPER salute us as we make our way in. The floor we visit looks like a warehouse, filled with a miscellany of objects that seem to have little to do with one another, except that they all want to be sold. Jewelry, old currency, lamps, a toy horse, a piano, a rug taped to the floor at the edges. My father loupes some items of jewelry, pressing the small magnifier as close to his left eye—his better one—as he can without scratching his glasses, while his right eye takes on a slackness, almost like a lazy eye.

We are not there for long, and before we leave, my father says the first Yiddish phrase I remember learning as a child, a phrase I once proudly repeated to the principal of my Jewish nursery school. *Gurnischt mit gurnischt.* Nothing with nothing.

When we get outside, my father tosses the auction brochure into the trash. He has a feeling that one of the jewelry pieces he saw has been modified, that a new stone has been put into its antique setting. This would mean the piece is not a complete original. My father tells me that auction houses make no guarantees on this aspect of the business, but it is important to an antiques dealer like him. Purchasing jewelry has become more complicated. When he first started, not everything he bought needed to be a prize piece. There was a flip market for cheaper merchandise. Over time, though, the customers who used to buy inexpensive goods have lost their spending money. Those with thicker wallets want only first-class items.

Finally, my father and I are on our way to see the fifteen carater and to try to sell the goods he's packed beneath his jacket. At the exchange, we find the middleman's booth. Instead of waiting on the outside of the showcase like other customers, my father squeezes into the tiny workspace and takes a seat. On the back wall hang pictures of the dealer's children and nephews and also one of the Lubavitcher rebbe, though this man is not a Hasid. A sign says *No Refunds Exchanges Only Within 7 Days.* Music plays on a stereo.

"Music?" asks my father.

The dealer at the exchange, a tan, bespectacled man with a Persian accent, says, yes, if it's okay with him.

"If we can dance," says my father.

The dealer starts to dance, moving his rib cage from side to side. Then he hands me an envelope with "California Lifeline Telephone Service" written on it. He jokes that it's for me. On the back of the envelope it says "Oltuski." At first I'm confused, but then I understand. What else can it be but a check?

My father looks at the Cape stone. He asks, "Who do I talk to about price?"

The exchange dealer gives him a number, and my father dials. He says "twelve" to the man on the other line. The man, the exchange dealer's selling partner, wants more. My father believes he should get a discount if he delivers a check immediately. He tells the dealer, who is in California, that he saw the needle—a flaw—in the stone right away. What he is saying is that the stone is not worth as much as the dealer is asking for it. I learn later that the needle is black and that Asian dealers do not buy diamonds with black inclusions, because the color signifies bad luck. This limits the market in which my father could resell the stone. Not only will he not be able to sell the stone to an Asian dealer, but he won't be able to sell it to dealers who want to be able to sell to Asian dealers. My father continues on the phone. At one point he says, "None of your business."

We leave the exchange with the stone but without an agreement. First, my father wants to show Roy. No one at the exchange knows it is Roy he's working with. This is my father's right, to keep his partner a secret. Many deals involve a mystery man. Sometimes a buyer keeps a hidden identity, because he's known to hold an expensive inventory and fears a seller will hike up his price if he finds out who the buyer is. Maybe buyer and seller just don't get along. Or maybe it's simply a desire for anonymity. Diamond men are such masters of secrecy that they basically speak a different language, separate from Yiddish and English—it is the language of discretion. In this language, there are no names or details or original prices. Once, while trying to buy a diamond on Forty-seventh Street, David Abraham, the former Club VP, asked the seller how much he paid for the stone. "My cost is nothing," said the man cryptically. "I found it on the street."

We walk along the sidewalk, parallel to the noisy congestion of Fifth Avenue, together with other hustling dealers, until we are a block away from Forty-seventh Street. We enter Roy's office near the secretary's quarters, which are sealed off in glass. It is a spacious workplace, with several meeting rooms—in one, a pair of foreign dealers waits for Roy—a gemologist's chamber, a four-man cutting factory, and a large main office with two desks: one for Roy, and one for his son Adam.

Roy's cutting factory was the first one I'd ever been to, and it introduced me to the dark, cramped world in which diamonds, bright and gleaming, are shaped. The balms and potions that grace the countertops, the recordlike cutting wheels, the scent of chemicals, the strange assortment of recycled containers (black spherical photo negative tubes, pill boxes) in which diamond dust and other small substances are kept, the staticy hum of stone upon wheel, the dusty floors and gray walls contrasted with the stark fluorescent diamond lights, like a construction site at night.

In Roy's luxurious office, with walls the color of sea foam, my father takes the fifteen carater out and spins it on the table like a dreidel. It lands on its side.

"It's ridiculous that something like this is worth so much money," he says to me.

Soon Roy comes in and looks at the stone. Today, there is little small talk. His opinion is that it has a deep bruise. Diamonds are a bit like human bodies. They, too, can be concussed, but while a person suffers blood loss, a stone will only crack, little veins of fractures stemming from the site of damage. My father and Roy talk about the diamond's shortcomings. Roy believes the stone will lose more weight in its cleanup than my father originally thought.

My father takes it into the cutting factory to consult with Roy's cutters. One of them, a Hasidic man, agrees that the stone will lose weight. Under a light, he examines the diamond for stress, interior lines that could make it shatter on the cutting wheel.

My father looks disappointed.

"What do you think it'll lose?" he asks the cutter. He clarifies that he wouldn't remove the gletz, the flaw visible from the outside of the stone.

The cutter tells my father that the gletz is really two gletzes.

Before we leave the room, my father asks the cutter for reassurance: "But the stone has life the way it is, right?" Life is that indefinable virtue of a diamond, after all its tangible qualities have been tallied up. It is the measurement of its temperament, and those who work in diamonds know that they have temperaments. Really, my father is asking the cutter if the diamond is pretty.

"Right," says the cutter.

When we come out, Roy, who has been with the foreign

dealers, asks, "Paul? Is it dangerous?" He wants a prognosis. He wants to know if their potential investment is likely to burst on the wheel. But the stone isn't dangerous.

After some back-and-forth, Roy and my father decide that although they will make an offer, the diamond is not worth too much money. Roy thinks it has an ugly make, that it's too flat. My father calls the stone's long-distance owner in California and offers him twelve thousand per carat. The man hangs up.

We go back to the exchange to return the diamond. The dealer tells my father that he is foolish for not accepting the deal, because just that morning, he had spoken to the seller, and the man said to him that this diamond, the Cape stone, was for Paul.

This catches my father's attention. He gets back on the phone with the Californian. He asks the man what the stone *has* to cost. He makes another offer. The dealer on the phone calls him cuckoo.

"Don't call me cuckoo."

Then the dealer at the exchange gets back on the phone with the Californian. My father nods yes at me. I have no idea what's going on, whether we are winning or losing, but I don't want it to stop.

The exchange dealer tells my father that they had decided to show him the stone before anyone else, that my father is making a mistake. Eventually, my father says, "I'm not upset if you say no, but I wouldn't pay a dollar more than twelve five."

The dealer tells my father that he makes, not breaks, deals; he calls California once again. My father says to the man on the phone that he appreciates that they called him first. Soon, he passes the phone back to the exchange dealer.

We wait. The sound of a steaming machine drifts over from another booth. We wait some more for the phone to ring. The dealer raises the volume on his stereo, then lowers it. He calls, "Sixty off," across the aisle to another man. The dealer in the other booth calls back, "Fifty off"—the back-and-forths of another deal.

Our dealer mentions a third man who says he paid fifty off.

"That's his problem, not mine."

Their debate is put on hold while my father and I run out to get a package from my mother on the street—another stone. She pulls the car up to the exchange, says hi, hands my father a small manila envelope, then, with a quick good-bye, drives away. She had probably driven him to work that morning. My parents aren't big fans of public transportation and end up haggling over who gets the car almost every day, my mother wanting it for errands and to take it to the West Side, where she practices piano at Juilliard. As a compromise, she's become a sort of default chauffeur for him, driving their Mercedes first to Forty-seventh Street and then on to wherever she needs to go, as though she were still living in Germany, which she has never stopped missing, flying across the autobahn at liberating speeds.

Back at the booth, the exchange dealer calls his long-distance selling partner and gives my father the phone. They talk. My father says that he already went above his limit. "Sweetheart, don't . . ." I can tell he is aggravated.

He has already given up on the deal, we have already left the exchange, when he gets a call on his cell phone from the Californian. We stand in the marble floored vestibule of another office building, and my father tells the dealer not to make it personal. But then he says to hold on. He looks at his phone, puts it back to his ear, and utters, "*Mazal* one ninety," $190,000—$12,667 per carat. He has been pushed up almost $700 per carat. I was not expecting him to give in, but he and Roy now own a fifteen-carat Cape diamond.

Mazal is the word that closes diamond deals all over the world. When two people say *Mazal*, short for *Mazal und brucha*—"luck and blessing" in Yiddish—the stone has transferred possession, no matter who is physically holding the gem or what other offers the seller gets. I've heard Indian dealers say it, Muslim dealers say it. It's just another way in which the ancient customs of Judaism run deeply through the diamond business.

Outside the elevator, I ask my father what he needs to sell the stone for. He shrugs, smiling. "Whatever I can get."

We ride the elevator up and go from office to office, my father unpacking the baggies in his chest pack and laying his jewelry out for other dealers to see. The first room we stop in is small. My father takes a seat at a table across from the dealer we're visiting. There is a necklace lying on the table, and my father asks how much. The man says my father can't afford it. He shouldn't even be seeing it. My father jokes that he'll tell everyone if the dealer doesn't give it to him cheap. Then he sets out the contents of his chest pack—an assortment of earrings and pendants from Tiffany & Co. The pieces are filled with round diamonds framed by thick, almost liquid-looking platinum, and embellished with the occasional bubblegum-pink sapphire.

My father gives him the jewelry's retail cost. Now it's the dealer's turn to be outraged by the price. He says Tiffany should be embarrassed. My father counters that if Tiffany didn't do what they do, he wouldn't survive. All Forty-seventh Street dealers are, in some way, dependent upon the more expensive stores higher up Fifth Avenue. These shops set the retail prices, which, in turn, impact wholesale prices.

The second dealer we see doesn't want the Tiffany pieces either, but he looks at a diamond my father shows him, pointing out that it has a chip.

"You know it's chipped," he says a second time.

My father tells the man he's slowly evolving into his father.

"If it doesn't hold three carats, will you give me a refund?" the dealer asks.

My father laughs. "*Make* it hold three carats." Though his tactics don't seem entirely convincing to me, he manages to sell the stone.

Our last stop is a social call to Hartley Brown, a Scottish Jewish dealer my father met at a show about twenty years ago. He is a tall man with the type of elegant brogue I believe exists only because I've heard him speak.

When we come in, Hartley's safe is open. His merchandise is spread on his table, and he sits before it like a little boy who has made a mess.

The office is like a jeweler's playground. Hartley shows us some of his treasures, such as a mechanical brooch with diamond-encrusted rods that spin in opposite directions beneath a crystal window (my father pretends to get hypnotized), a bird-shaped pin with rubies for eyes, and a necklace he designed, made of rough diamonds, platinum, and prehistoric fossils imprinted with the form of small beetle-like animals. He tells us a story about a peridot gem that was dropped on a carpet at such an angle that it was cleaved along its grain, and then shows off some earrings that he and his wife made out of colorful clusters of gems and diamonds. He produces a gemstone ring and shines his natural daylight lamp upon it. He asks me what color the gemstone is. I say green. Then he shines a flashlight on it. This time I say brown or orange. The stone is an Alexandrite, prone to color change because it contains three tones beneath its surface, each of which responds to different types of light.

My father says we have to go, that we got *Mazal* a while earlier, but before we manage to leave, Hartley has to show us a sapphire that changes colors and a ring with a carved stone that dates back to the Roman Empire. Finally, we make our exit.

On our way back to the exchange, my father says, "I bought something, and I sold something today." After all, that is what the business is. If a dealer doesn't hold on to any one diamond for too long, he is doing all right.

At the exchange, the dealer at the booth reminds my father of his loyalty to him. My father reaches over the showcase, puts a hand on the side of his neck, pulls the dealer toward him, and kisses him on the cheek.

This was the first time I'd witnessed my father really barter. Usually, he speaks of "aggressive bargaining" with a certain disdain, as though it were a primitive habit, like catcalls and bar fights. He can't be bothered with all the shouting, the ever-converging extremities of price ("Ten thousand." "Nine thousand." "Nine seven." "Deal"). Some dealers—I've noticed a trend among sellers of antique jewelry—pride themselves on the very fact that they *don't* care for true haggling.

But other Forty-seventh Streeters engage in the ritual with appetite. Bargaining is a form of entertainment, a performance in addition to a way of arriving at price. Like characters on a stage, dealers are often watched by others when they go at it, and, to further their cause, they choose from a rich arsenal of bargaining techniques.

There is the panic method, where the buyer reminds the seller of how slow business is, how no stones are selling, then comes in with his offer. Then there is the sympathy card. Some dealers have even brought children along with them to a deal, presumably to make the need of their income more palpable, or simply because it's more difficult to say no to someone in front of their kids.

Others go about standing their ground more quietly. But even those buyers like my father who ask, "What is your lowest price?" and expect to be answered truthfully are really participating in at least one round of negotiation. The no-nonsense question is itself an approach to bargaining.

Only rarely will a buyer make a flat-out offer upon seeing the stone, before even asking, How much? "That's a very adventurous buyer," David Abraham said to me, "right? That's one in a million. Okay? Those are legendary kinds of people that know how to buy, all right?"

More often, buyer and seller will ask each other all sorts of questions, feel each other out with words and numbers. David calls it a "cat-and-mouse kind of a game. You're pulling the cheese away. *I'll give you the best price I can,* for example. *I'll give you ten thousand, that's really the best I can do.* But really, he can give a little more." Just as the seller can, in most circumstances, take a little less.

Mr. Rawat, a tenth-generation dealer from Jaipur, told me that Indians take pleasure in bartering. "If you tell them hundred, they will not like to take hundred. They would like to bargain. It's not like fixed price."

But Mr. Rawat's son, Neeraj, has suffered from the opposite stereotype. He told me that some people assume he is less forceful because he is Indian and can be bargained down more easily. The first time Neeraj had ever been sent out as his father's employee, the man he was trying to sell to derided him. "This guy is like, *Oh, you Indians, you this, you that.* I just wanted to tell him off, but I couldn't. I was better educated than him."

Bargaining can be mean. It can be nasty and unpleasant. And hypothetically, it can involve a lie. When a dealer takes a stone on consignment, the stone's owner writes down an amount on the receipt. The borrower may sell the stone if he finds a buyer willing to pay that amount—say, a hundred thousand dollars—or higher, or he may keep it if he himself decides to pay the price. Oftentimes, while he is in possession of the gem, the borrower will try to negotiate the price with the owner. Perhaps he has another party interested in it, but only for ninety-five thousand. If he sells for ninety-five, he'll be losing money. But if he gets the owner to come down to $92,150, he can still make a three percent commission. He might speak frankly with the owner and tell him that he has serious interest for ninety-five thousand, then ask if the owner can come down a bit. The owner has to trust that the borrower's third party really *is* demanding ninety-five thousand, but this is not always true. The borrower might be trying to make a buck. He might have an offer, or he might even have sold the stone, for a hundred thousand dollars, but hoping that he can make an even nicer profit, he tells the owner that his buyer was insisting on ninety-five.

This practice, my father professes, is "for the super sly." Chris Morris, an antique jewelry dealer in the district, has wondered about some of the people with whom he's done business. "I might give a stone to someone, and they wanna show it to a client of theirs and then they call me back and they tell me that it's not sold, but can I do better on the price? And in my mind, I'm sitting there saying, *Well, I know they sold the stone,* and it goes back and forth. Do I believe that they've not sold my stone and they're just trying to get more money off of me, or vice versa?"

There is another danger to claiming a consignment diamond isn't sold when it is: its owner can demand the item back. Perhaps he has a better offer. Perhaps he's fed up with the way the deal is going. Perhaps he's changed his mind. If the buyer has already sold the stone, he chances owing the owner a diamond that is no longer in his possession.

Consignment can bring with it a whole world of chaos, not just for the man borrowing the stone on memo but for its owner, as well. Efraim Reiss, the son of a diamond factory owner in the district, told me about a three-carat brown diamond he once gave out on consignment to a large company. After a week, the company returned the stone unsold. Next, Efraim gave the diamond to my father, also on consignment. Right after my father acquired the stone, the company called Efraim back. "'I have the stone sold. *Get me back the stone.*'"

"The stone was by your father," Efraim told me. "I called him up. 'Paul, I need the stone back. I have the stone sold. You know, it's difficult times today. I have a sale, I wanna sell it.' It was on Purim, so your father says, 'Call me tomorrow. I just really gave out the stone two days ago. It's a very good customer. I don't want to get it back so fast. Call me back tomorrow, I'll see what I could do.'" Now Efraim was in a bind. He didn't want to rush my father after only a few days with his diamond, but he didn't want to lose his opportunity to sell the stone. The chance might not come again.

"So I called back your father," Efraim told me, "and he was so nice, that he said, 'You know what? How much do you have on the stone? I'll buy the stone.'" Efraim even found a second stone for his other customer.

Keeping your adversary happy during bargaining is important. People have different ways of doing this. There are smooth talkers who praise the dealer on the other end, while trying to persuade him. Once in a while, David Abraham will flip a coin if he and a dealer are down to a low margin.

"Why?" said David by way of explanation. "Because the hundred dollars means nothing. The important thing is that you have a relationship with him, okay? So if you win, right? Let's say you win, as a buyer, and it has to be ninety-nine hundred, so you wanna make the other guy happy, say, 'You know what? I won, but anyway, I like the stone, and I don't care about the thing, so let's make it ten thousand.' *This guy*, with that hundred dollars you just gave him, from the bet he just lost, that guy will give you a thousand dollars back" in future deals. I told David that was very smart. Egged on, he offered me another trick. "You wanna make somebody happy?"

"Yeah," I said.

"I'll give you another way. It's so dumb. Let's say your whole total is ninety-nine thirty. Total, all right? Okay? You make a calculation. You owe me ninety-nine thirty, say, 'You know what? Make it ninety-nine fifty, 'cause I don't like the small amount in the middle or whatever,' or make it ten thousand even, right? For seventy dollars, you have just gotten seven hundred dollars' worth of goodwill that you can use later, 'cause when you're gonna buy something from the guy next time, believe me, you can slaughter him, you know? And he's gonna say, 'What? Why such a bad price?' You're gonna say, 'What are you talking about? You're my good friend, who cares about money with you?'" David paused and assumed a pedagogical tone. "I've always learned that if you give people what they want—I'm not talking about prices, by the way—if you give people what they want, it really goes a long way. A big no-no in the diamond business," he added, "is to buy from somebody very cheap." If a dealer knows that his colleague is desperate to sell, he can force the man into giving him the diamond at an extremely low price. "Don't hurt them too much," said David benevolently.

Dealers are always talking about relationships, as though they couple up with each other and stride earnestly off to therapy. They casually mention things that are good or bad for the relationship, or that "strain" the relationship, such as the failing economy. "Some partnerships work like a marriage," David told me, "very easily and very well, and some don't work out so well." Others, he compared to a fling.

When one of his friends, a man he had once partnered up with on some merchandise, stumbled upon our conversation in the Diamond Dealers Club, David said, "See, in diamonds, you can have like a brief romance, you can have a friendship for a month, a week, a day, a year, whatever. Ten years."

Some dealers spend a tremendous amount of time with one another. They share an office. They lunch together. They travel together. They keep each other's secrets. When I asked an Iranian dealer what his favorite thing about the industry and Forty-seventh Street was, he answered that it was the friends he'd made, the close relationships he had formed during his almost thirty years of dealing. This, he said, he valued "much more than the diamonds."

But even friends are not exempt from bargaining. Negotiation can be pleasurable, full of compliments and stories. David used to sell diamonds to a woman in Bangkok who called him "sweet mouth." In Thai, David said, "it's called *Pakwan,* which means 'sweet mouth.' Right? She says that your mouth is sweet, so she likes to hear the stories. They don't care for the diamonds as much as they like the person."

In the middle of his story, David said hello to another Club member who had materialized at our table in the small café at the back of the hall. The man told me he had sold a diamond to my grandfather the week before. David filled him in on our bargaining conversation. Soon, talk turned to Elvis, the singing gem dealer.

"You know Elvis?" David asked the man. "He tells stories, he makes songs. He turns people on. You can do business like that."

It was because of my grandfather, Opa Yankel, that I first met Elvis. Opa thought I didn't wear enough jewelry, so he took me to the Diamond Dealers Club, where he spends most of his days, to outfit me. There I was, standing near the station where dealers go to weigh their diamonds, acquiring a sapphire bracelet and earrings from a fifty-something-year-old Jewish Elvis impersonator wearing designer clothing and a diamond-encrusted charm necklace. The metal and diamond charms spelled TCB, for *Taking Care of Business.* Upon my grandfather's command, Elvis had pulled the chain out from underneath his shirt.

Elvis is a diamond and colored-gem merchant who can be found hanging around the Club most afternoons. He often breaks out into song, dancing Elvis-style, with one leg stationary and the other flexible as jelly. Sometimes you can see the lines left by the comb that went through his pompadour that morning. His sideburns are not quite as prominent as Presley's. His accent is half Israeli, half Bukharan, but all of that disappears when he sings. He really does sound like the King. Among all the characters who roam the street, he is the most colorful.

"You could say I'm like a woman when it comes to diamonds," Elvis once told me. He can't get enough of them. A fan of fine garments, he dresses in silk shirts and lacquered alligator shoes and keeps close track of sample sales throughout the city. I'd never met anyone like him, and I knew from the beginning that I never would again. Spending time with him is like spending time with a mime who wears his makeup during off-hours. I can usually count on a serenade when I run into him in the district, and at the end of a phone conversation, he signs off with "Daddy loves you."

The day I came to see him with Opa, Elvis showed me some bracelet-and-earring sets, mostly rubies or sapphires surrounded by diamonds in the shape of a flower. As far as I could tell, there were four options: rubies with white or yellow gold, or sapphires with white or yellow gold. I picked the sapphire and white-gold set. My grandfather named a price, which Elvis rejected. To emphasize his resolve, Opa wrote the number down on a piece of paper. Elvis said no, he was sorry. He said he didn't even want to give Opa the idea that he would sell for that amount. He said he respects Opa so much. My grandfather is not a big talker. He underlined the number. Elvis walked away, though he couldn't go very far in the Club, which is composed of two main rooms, a restaurant, a chapel, and a few smaller lounges.

I thought the deal was dead, but shortly afterward Elvis returned with the items. All this is part of bargaining. Sometimes a dealer may have to laugh at an offer in order to indicate how absurd it is. Or he might raise his voice; offers have the power to offend. Once, a broker was so floored by a number that he recited Kaddish, the Jewish prayer for the dead. My grandfather isn't in show business. On occasion, he will raise a hand in protest, but that is about it. He simply states his prices and saunters away. When Elvis returned, he agreed to Opa's price.

I felt uncomfortable watching my grandfather's firmness triumph over Elvis's pleading antics, but Elvis didn't take it personally. He couldn't find enough words to praise Opa. He told me that he looks up to him, "because he reminds me of the movie stars, you know the old movie stars, the very special ones, like Clark Gable or like Paul Muni, or Errol Flynn, or Tyrone Power. He's a powerful man. He's not like a woman. You know? He's a strong man, like John Wayne." Although the image of a Yiddish John Wayne is not easily conjurable, I know what Elvis means. My grandfather dresses in day suits and fedoras; he is very tall, and the fact that his legs are of slightly unequal lengths results in a kind of sauntering that, combined with his upright posture, implies a deep-seated confidence.

Opa Yankel's style of bargaining is like his walk: firm and self-assured. He is respected on Forty-seventh Street for intuiting how much he can get out of a person, but also knowing when to yield for the sake of peace and goodwill. This I learned from Jack Reiss, Efraim Reiss's father.

Mr. Reiss spoke of my grandfather's bargaining instinct as a total-body experience. Opa, he said, knows how to look at someone's nose. "You have to have the sixth sense, like they say. It's not only the diamond. He knows if the guy is stronger or weaker, and he feels he can get a better price, and you have to be smart for that and you have to really have sharpened your ears and that's something that is the old, old school. As tough, as smart as they come, but *good.* And he knows when to give, you know what I mean? See, the difference between today, the young guys, sometimes, they don't know when to give. Sometimes you gotta give a little bit. To make the deal . . ."

The most romantic rite of bargaining is the cachet; even its name suggests a thrilling secrecy. The cachet is used when a broker negotiates the sale of a diamond. He approaches dealers with the stone on behalf of its owner, and if a dealer is interested enough, he makes an offer. To assure that the broker does not show the stone around town, the potential buyer puts the diamond in a small manila envelope. This is the cachet. The buyer seals the envelope and, on its outside, writes his offer, along with how soon he will pay, and whether by cash or check. Then the broker takes the stone, in the cachet, back to its seller.

Even the owner is not allowed to open the cachet that holds his stone. He has three choices. He can accept the buyer's offer, stand his ground on the original figure, or make a new offer. Either way, he must send the broker—and the cachet—back to the buyer with a chance for purchase. The buyer can either accept with a *Mazal,* put a line through his previous offer and write in a higher one, or bow out of the deal by opening the cachet, thereby releasing the diamond back into the market and freeing the broker to show it to others. Sometimes theatrics kick in. If his offer has been declined, the buyer might slowly tear a corner of the envelope, to show the broker he's serious about pulling out of the transaction. Sometimes the broker will stop him, take the envelope back to the owner, and ask him to reconsider the buyer's last offer once again, since they may be losing him.

The cachet is an old tradition. It cuts to the core of the diamond business, veiling a gem inside a vessel as bland and mediocre as a manila envelope, carefully delineating the boundaries between private and public, between that which is hidden and that which may be revealed.

Chapter 4

Carbon

The path of a diamond, from the mine to its eventual customer, can span continents and involve countless people, destroying some of them and saving others. It employs almost every faculty mankind has mastered—art, science, commerce, and force. And it always has consequences for someone, somewhere.

One hundred miles below the surface of the planet, in the earth's upper mantle, the temperature seethes at over two thousand degrees Fahrenheit, and the pressure is as forceful as tens of thousands of atmospheres. At least a billion years ago, these circumstances drove underground carbon atoms together into the strongest kind of bond: the shared electron bond. Carbon at its densest is a diamond.

Scientists think that at least some of the mantle carbon from which diamonds were born came from small gemstones that got to the earth via meteorites. After that, carbon may have continued to grow around these ur-diamonds, making larger ones. Once grown, the diamonds sat waiting in the upper mantle until the pressure on magma deep in the earth caused it to explode upward, forcing itself through the cracks of rocks, sometimes through rocks themselves. As the magma rose, it swept up anything it met along the way, so when it passed through that sultry area in which the diamonds cooked, it seized them. On its way to the top, the magma quickened—pressure let off, as less ground pushed down upon it—then blasted upward at one hundred miles an hour.

At first, the rising lava, along with its spoils, took the shape of a carrot. But as the mixture rose, it grew wider, into the shape of a bowl. Close to the earth's surface, the magma became partially gaseous because of the dwindling pressure. And when nothing much held it back, it burst through the ground, bringing with it the rocks, the gases, and the diamonds.

Soon, the lava crystallized into a rock called kimberlite (its namesake is the city of Kimberley), which eventually eroded. Layers of new earth concealed it, and some surface stones, like the one Erasmus Jacobs found in South Africa, were carried away from their original source via streams and rivers. The others rested below the earth in this pipe of diamonds for millennia.

A diamond is not guaranteed to reach the top. In fact, that's rare. Often the magma doesn't travel fast enough and exposes the diamonds it's carrying to regions lacking in sufficient pressure or temperature. If a diamond lingers too long in such regions, it turns to graphite. But even those stones that, against all geological odds, make it to the top unharmed are only at the beginning of a long pilgrimage that will bring them to their customers.

✡

About fifteen percent of the world's gem-quality diamonds are unearthed by human hands. Hunched over a few austere tools, up to their waists in dirty water in the roasting heat of Africa and South America, more than a million artisanal diggers spend as much as ten hours a day searching for diamonds. After burrowing through gravel near the waterfronts, the diggers bring sieving pans full of sediment into the water, submerge them, and rub their palms against the hard grains to separate the ore from the gemstones.

Payment varies. Generally, it includes a meager lunch, about a dollar per day, and, depending on the site, possibly a small portion of the gemstones. Sometimes there are cave-ins. Sometimes there are parasites in the water. Often the workforce includes children. Many of the diggers will never see a polished ready-to-wear diamond in their lifetime.

Artisanal mining belongs to the larger category of alluvial mining, the extraction of diamonds that have floated to deposits from their kimberlitic origins via water, even water that has since dried up. Because many of the diggers operate beyond the regulatory parameters set up for the international diamond trade, artisanal digging can destroy surrounding land, making life dangerous for the rest of the local community. One of the biggest problems is that artisanal mining leaves pits, which can fill with water that is stagnant and is a breeding ground for insects carrying diseases, among them malaria and dengue fever.

At the same time, artisanal mining employs about 1.5 million people and sustains approximately 15 million people worldwide, including diggers' families and communities. These diamonds are usually sold by leaders of artisanal mining rings to individual buyers or companies.

Of course, there are larger ventures that have the capital

to safeguard and improve diamond areas. Target Resources, a company that supplies Tiffany & Co., constructed sewers, schools, and roads in Sierra Leone. Tiffany itself helped create a project to cover the mining pits that artisanal diggers created. And in 2007, Nicky Oppenheimer, De Beers' chairman, and his wife were given the WWF-Lonmin Award for environmental conservation.

Today, De Beers sells about forty percent of the world's rough diamonds and owns or co-owns more than fifteen mines in Africa, some together with the governments of Botswana and Namibia. Mines exist not only underground but also on river banks and under the sea. Depending on the setting, De Beers recovers the stones with explosives, track dozers, ships, hoses, divers, even special machines created to stalk the seabed, collecting material in which the gemstones reside.

After they're extracted, De Beers diamonds travel to one of four sorting offices in Africa and London. First, the sorters classify them by weight, then, into "sawable" rough diamonds (which eventually get split into two diamonds) or "makeable" rough diamonds (from which only one stone will emerge). Next, they inspect the diamonds for flaws. This is when a stone's fate is determined—whether it will become a jewelry diamond or an industrial diamond, used in machines or to cut tough materials such as metals, teeth in a dental office, and other diamonds. After this, the sorters separate the stones by color. By the end of the whole ordeal, there are about twelve thousand categories of diamonds. Each category corresponds to a value. In this way, sorting diamonds is also how De Beers and the mining countries settle on pricing for their transactions. Both sides must approve the stones' groupings. In the event of a disagreement, they can use a collection of sample diamonds for comparison, accepted by both parties as standard. After they've reached a consensus, the diamonds become the property of the Diamond Trading Company.

The DTC then aggregates its newly purchased stones, grouping them into mixed clusters of various sizes, colors, and qualities. These diamonds are then sold to sightholders or sent to De Beers stores and dispersed throughout the world. The aggregation is commonly done in London, but in 2008, the company announced that it would be relocating the work to Botswana. This was part of a push during the first years of the century to move the value-adding portions of the mining process—those that come after the actual extraction—to the African gem-producing countries. These countries could then reap greater benefits from the resources that are coming from their own mineral deposits.

Botswana is the industry's shining paradigm of an African country helped by diamonds. De Beers first uncovered a kimberlite pipe there in 1967. By 2008, Debswana, the company owned jointly by De Beers and the government of Botswana, turned out the highest amount, in worth, of diamonds worldwide. There has even been talk of Botswana as a possible competitor to New York, Antwerp, and Tel Aviv—the trade's historical manufacturing hubs—for large diamond cutting. "The diamond industry's center of gravity is shifting. . . ." said De Beers chairman Nicky Oppenheimer. However, because of the economic recession, De Beers deferred the scheduled move.

I think of De Beers as a godlike presence in the diamond world—a distant, looming power that controls from afar. Its sightholders are high priests. Like God, De Beers functions with many names: the Diamond Trading Company, the Cartel, and, in the old days, the Syndicate.

Although more than ninety percent of diamonds that come to America pass through the district, very few Forty-seventh Streeters have a sight with De Beers. Most dealers there are so far down the food chain, they're lucky to be third, fourth, or even fifth in line to get their hands on stones. "We are like little popcorns," Elvis told me. "We are nothing. We are just survivors for pennies here, pennies there. *Lechem le'echol, beged lilbosh,*" he quoted me from the Hebrew Bible, an oath Jacob took. *Food to eat and clothing to wear.* "Just to survive."

The retail stores farther up Fifth or Madison Avenues belong to a diamond world more lavish than Forty-seventh Street, a world full of familiar colors, logos, and names. These stores even look at gems and jewelry differently. "Up at Tiffany they would just sit there and measure stones, and if they had *any* black inclusions, they wouldn't accept it," said Chris Morris, who spent a year grading diamonds at Tiffany before becoming an antiques dealer. When he relocated to Forty-seventh Street, he learned that "you just had to open your eyes more and look at the total beauty of the stone overall, whereas they look at a stone and it's more of a piece of paper."

Moving to the district, Chris began to understand the anxiety that came along with being a dealer. Would he be paid or stiffed for the necklace he had just given away on credit? Would he get safely from his office to his colleague's, or from New York City to the other cities in which he dealt, with hundreds of thousands of dollars (recently a piece for \$1.4 million) strapped to his body? To show me where he kept his goods, Chris stretched out in his chair, exposing his midriff. He slid his hand into his jeans and pulled out a gray pouch with a belt. "I dress understated. No one thinks that a bum like me is gonna be carrying around a lot of merchandise." He wore a sweatshirt and looked like a man for whom a jog outside is never far away.

Some of the fancy-store vendors look down upon the diamond district and its dealers as inferior members of the profession. Years ago, a saleswoman at a small shop on Fifth Avenue told me how Forty-seventh Street was like a third-world country. She likened it to Cambodia, many of whose citizens, she reminded me, lacked toilet paper. At Tiffany, Chris told me, "it was sort of inbred that we were Tiffany, that we were the best, and what they sell on Forty-seventh Street is a much lower quality, but then the reality is, when you come down here, you realize that they were just putting on a big salesmanship game." In the district, he learned to get acquainted with real people's budgets. Sometimes that meant selling them a slightly lower-quality stone for less money. He started asking them what they were looking to spend instead of convincing them to forfeit two months' salary. Not everyone needs a D flawless.

Despite their differences, though, there is a lot of interaction between Forty-seventh Street and the more expensive stores. My father used to do business with Fred Leighton, one of the most eminent jewelers in the country. The company was later sold and went Chapter 11, but back in the seventies through the early 2000s, Leighton was the king of estate jewelry, with pieces from nearly every historical period—Victorian, Edwardian, the twenties, the thirties, the forties, the fifties, art deco. My father remembers a time when "dealers from all over the world would make him their first stop," when movie stars and personalities would be taken into the back room of the Fred Leighton store on

Madison Avenue for lobster lunches with caviar and champagne on Saturday afternoons. To my father and his friends, Leighton himself was a kind of star. I remember going on walks with my grandfather, and if we were close to Leighton's shop, we would stop and peek into the windows so Opa could see the kinds of pieces that made a jewelry man famous, and in retrospect, I think I may have seen a tinge of envy in his gaze.

One day my father sent my mother over to the shop with a large diamond necklace. My mother took the necklace into the Madison Avenue store with the rounded awning and the bronze-colored storefront name. Fred Leighton, whose real name is Murray Mondschein, attended to her himself. "I was very impressed with him," she says to me. "He was so charming, and he was so . . ." She breathes in. "You know he gave me the feeling like I sold him the piece, while I was more or less the messenger for Daddy, you know?" The measurements of the necklace's gems had already been taken, a price had even been hammered out between the jeweler and my father. Still, Murray made my mother feel as though she herself had done the deal. When she was in the store, he focused only on her and made her believe that, as my mother puts it, "he would have taken anything I showed him."

Whether they sell in an office the size of a closet or a shop with window displays as high as the heavens, all diamond vendors use the same elaborate set of criteria to judge the stones they buy and sell.

After being cut and polished, a diamond usually makes one last stop before it's ready for sale. In the laboratories of the Gemological Institute of America, the stones are measured, scrutinized, and given a report according to the Four Cs of diamonds, a little set of commandments that the GIA established and to which every dealer subscribes: carat weight, color, clarity, and cut.

When a diamond first arrives at the GIA, it gets a bar code and its weight is noted to the fifth decimal point. A computer makes a record of its dimensions and the angles between its facets, the tiny windows surrounding the diamond. Then the stone travels around the laboratory in a clear box containing only its ID and data, with none of its owner's information, for anonymity.

First, the lab grades a diamond for color. When laypeople see a white diamond, they see only white. Maybe they'll see a yellowish tone, but that's about it. But dealers know that diamonds can fall into seventeen colors, from D (perfectly colorless) to Z (light yellow or brown). Then there are the special colors—red, blue, pink, purple, green, orange. There are also less desirable colors—slightly tinted yellow and brown, though brown has become more fashionable. And there are colors that would be less desirable but are so vivid that they *become* desirable, like deep brown or bright yellow.

Most diamond colors, whether sought after or not, owe their hues to imperfections and radiation. As the carbon layers of a diamond grow, small amounts of foreign material make their way between its tiers. These materials block certain colors of the spectrum. The colors that aren't obstructed shine through and give the diamond its tone. Nitrogen creates yellow diamonds. Boron delivers blue. And sometimes a glitch in the tiny structure of atoms within a diamond will turn out a brown, red, pink, or purple stone.

In a room with special lighting—all GIA labs have standard lighting and color—a grader meticulously considers a diamond's color in relation to a master set of diamonds whose colors are known, and decides which is closest. Then a second grader assesses the stone, blind to his predecessor's grading. If the two come up with different colors, and the stone in question is particularly large or of high quality, they call in a third grader to give his opinion. Sometimes they use a color-grading tool.

Then the diamond moves on to a "preliminary grader" who uses tenfold magnification to judge clarity. The GIA euphemistically calls a diamond's flaws "clarity characteristics," "inclusions," even "unique birthmarks," but, in truth, they are unwanted. As a diamond develops in the earth, it can be impregnated with a tiny crystal, which itself begins to grow, sometimes in abnormal shapes. If the crystal is visible to the grader, it's branded as an inclusion. A flaw on the outside of a diamond is a blemish. The preliminary grader chooses from eleven clarity scores, from F (Flawless) to I3 (Included 3), and that grade, just like the color grade, can determine the diamond's fate in the market. The grader makes a map of the stone's flaws on a computerized image of the diamond. He examines the stone's bottom and its girdle, the line that separates its upper muffinlike and lower conelike sections. If the stone is a "round brilliant"—the most common cut of diamond, with fifty-eight facets—he will also give it a Cut Grade. When jewelers say "cut" to one another, they're talking about shape: a tear-shaped "pear," a rectangular "Princess," a "cushion." But when GIA graders score cut, they look at a stone's proportions and finish, its craftsmanship. A diamond cut's real importance is how well the stone's facets play with light. A diamond glitters so colorfully because it's dense. It slows down the speed of light passing through it, cracks the color spectrum into different shades, and liberates them. Inside the diamond, light interacts with the stone's facets, like a ball bouncing off the walls of a billiard table. When a cutter cuts just the right number of facets in just the right places, a light show erupts within the diamond, like thousands of lightbulbs winking at you through the surface.

To the GIA, each of these little feats is a vital, carefully defined function. "Brightness" refers to the light a diamond reflects; "fire" is the stone's ability to produce a rainbow from light. "Scintillation" doesn't just mean shine; rather, it gauges the *amount* of glint produced when a diamond is moved back and forth. Privates (the industry term for a private customer) have other names for it—shine, shimmer, sparkle—as though diamonds were in a category with makeup and hair spray, and for some people, they are.

The preliminary grader's other task is to check the stone for any man-made treatments—the equivalent of plastic surgery on a diamond. This is something the GIA takes very seriously and investigates multiple times. An unmodified stone is more valuable than a treated diamond of the same size and quality. But in the avid hunt for the unnatural lies something of a contradiction: the very operations—a reduced imperfection, an improved color—that enhance a diamond from its natural state make it less perfect by industry standards, which means less expensive. In fairness, some treatments, like filling a diamond's minute crevices with leaded glass, can wear off, and so there is a danger in grading a diamond in its remedied state. However, in cases of permanent enrichment, as most color treatments are, I can't understand the drawback. Shouldn't a stone that looks like a D color and will continue looking like a D color for all eternity be called a D color? Not for dealers and jewelers. For them, these rules are gospel.

Not to say that there isn't a market for treated stones. Many customers are happy to pay a discount price for a diamond that looks flawless and colorless, even if it got to its current state with the help of technology. The GIA even has special grading standards for permanently color-treated stones. The rest of the natural diamonds go to a second grader, who reexamines them for treatments in the "double process step," just to be sure. This grader repeats the prelim's examinations, and when he is through, he views the prelim's markings to confirm his work. If the customer desires, an engravement can be lasered into the diamond, for the purposes of identification, branding, or sentimentality. Then the GIA prints up its report, puts the diamond back into its original packaging, and returns the stone to its owner.

The lab's method of evaluation is a highly scientific process that scrutinizes almost every one of a diamond's material qualities for a completely unscientific purpose: beauty. But in a way, that's what everyone in the business does—my father, Roy, Elvis, and all the other dealers. GIA reports are not the be-all and end-all of a stone's worth, but they occupy an important role in the buying and selling of diamonds. One of my tasks when I worked for my father was to keep his "certs" filed neatly in order of carat weight. He tried to have five copies of each available at all times, including when he was away at a trade show, which meant that, in advance of traveling, he and my mother spent hours xeroxing these papers to hand out to interested buyers when they stopped by my father's booth.

Certs offer dealers protection in bargaining. Without one, a buyer can try to call a Slightly Included diamond merely Included in order to pay less money, and a seller can call his Very Slightly Included diamond a Very, Very Slightly Included diamond to make more. My father is suspicious: "Without a cert, the buyer will automatically downgrade it." He refuses to buy diamonds above K color and I3 clarity without a report, period. The terms sound minute, and they represent near invisibilities, but to dealers, they mean thousands of dollars. Having a cert can increase a stone's value. And so, each year, thousands of merchants pay the GIA anywhere from fifty-three dollars to just under twenty-five hundred every time they submit a diamond. Then, armed with a report card, the dealers try to make a sale.

A young man came to see my father about a ring. At first I thought we were involved in an engagement, but as it turned out, the private had already secured his wife; this was just a gift. Private customers constantly wash up in the diamond district, delighted and confused, but they don't always make their way into the upper offices of Forty-seventh Street.

Unlike dealers who buy to sell, privates buy to wear. They are uncomfortable touching the stones at first. They worry about breakage, despite the fact that diamonds are the hardest substance known to man. They need to talk things over with their wives or husbands. Privates come with the notion that diamonds are forever, and for privates, who will not sell a piece of jewelry unless they're facing bankruptcy, they usually are.

By and large, my father sells to other wholesale dealers or

to stores. His goods are mostly secondhand, usually antique, and have made their way to his safe from the estates of dead rich women or recently ruined families through attorneys, auction houses, and other dealers. My father goes through many channels, sometimes traveling far, to scrape up the treasures of the dead, the indebted, and the absentminded.

More than once, he has spotted a jewel that he sold in another man's collection or at auction. Divorces happen, death. Pieces reenter the market. On occasion he has even come across a woman wearing his jewelry, but only those items he sold directly to another dealer, or a friend in Germany. Still, he feels a sense of pride. After all, these are pieces he handpicked, pieces he bought precisely because he thought someone else would find them beautiful.

Most of the time, wholesalers do not buy with a particular customer in mind. Their industry is backed not by the immediate reality of a woman but rather by the potential of that woman somewhere down the line. My father and his friends joke that the jewelry they trade never actually sees what he calls the light of retail.

But when the gems are in his hands, they enjoy an attention almost as lavish as they will receive being worn and fawned over by girls at a party. In his office, my father weighs them, pores over their dimensions, and takes their pictures. He will bring a diamond to his mouth, breathe on it, and wipe it clean. He will polish his rings with a child's toothbrush. It is as though he wants to make the experience pleasurable for the stones as well.

The jewels transform under his care. Sometimes he removes a stone from a piece of mediocre jewelry and has it recut and reshaped to bring it closer to perfection, or at least salability. His specialty is a particular cut called cushion, which is a square or rectangle with rounded corners. In diamonds, everything depends on the most minuscule of proportions.

In preparation for the private's arrival, my father arranged a few shallow trays of assorted rings. These trays are small square boards, covered with a black velvetlike material. They are pretty but not fancy. My father's office is certainly no Tiffany & Co. or Cartier. Unlike those posh shops, it is starkly lit by fluorescent lighting that makes you dizzy after a full day's work. Its gray carpet is hard and thin. In general, jewelry tends to look just a bit dingier on Forty-seventh Street. Instead of blue velveteen drawstring sacks, bracelets are packed into plastic baggies, rings come in recycled boxes, and stones in crumpled parcel paper, that blue tissuelike material enveloped in white. Although some pieces are worth hundreds of thousands of dollars, there are still reminders of Forty-seventh Street's roughness. Particles of dust that cling to the jewelry trays' velvet. Crossed-out prices on older items.

Price is impossible to ignore. Woven through the narrow clasps of necklaces, looped through the undersides of rings, bandaged onto the backs of diamond boxes are white price tags. Decades ago, when he first moved to America from Frankfurt, my father developed a secret pricing code. He took a nine-letter German word and assigned the first letter the digit one; the second, the digit two; and so on. On each price tag is written a jumble of letters that tells him what his original buying price was for the piece. This way, when bargaining starts, he knows how far he can go without losing money.

The private appeared on the small surveillance video, and my father went to greet him personally at the door after buzzing him in. For privates, you still have to put on a show.

They shook hands and sat at the desk in the front of the office, with the trays between them. Privates need to feel comfortable, taken care of, so my father made small talk. The man told my father that he proposed to his wife on Lake Placid, as though he needed to prove his love before my father would sell him a ring for his wife. Eavesdropping, I learned that she worked in the city but they lived on Long Island. He struck me as a romantic. Privates emerge from the outside world with notions of romance and love. But love is not on my father's mind when he tries to sell a diamond.

Like all dealers on Forty-seventh Street, he aims to make a profit. And so, in almost every transaction where the price has not been set, there will always be a moment when buyer and seller want opposite things, no matter how much they like each other, or how lovely the woman on whose finger the diamond will rest.

The private seemed nervous. A stone's qualities and defects are virtually microscopic and usually go unnoticed by outsiders to the trade, so many privates fear they are about to be ripped off. It's confusing to be told that a piece of rock costs as much as your rent or college tuition. It is confusing that diamonds are more expensive when they are the color of emeralds, which are often less expensive than diamonds. And it is confusing that fluorescence, which causes diamonds to appear blue in certain lights, is considered a shortcoming, but that stones from the Golconda Diamond Mine in India, which sometimes appear fluorescent, are the industry's holy grail.

The mood is different when it's dealers who are visiting. I was in the office once when a group of foreign diamond men came to see my father and his merchandise. They made themselves comfortable in his leather 1940s reproduction chairs. One crossed his legs widely. The men picked up various pieces. A dealer draped a slender bracelet around his broad, hairy wrist, holding the ends together.

"What is that one?" someone asked him.

"That's French." My father dug his hand into the tray. "That's French and that's French and that's French," he said, pulling out everything French he could find. When a piece caught one of the dealers' eyes, my father leaned in close as if they were going to kiss, and offered his loupe. The jeweler took the glass into his hands, while it still dangled from my father's neck, and looked at the piece magnified ten times.

Jewelers buy with their eyes. Most know fairly quickly if they want to purchase a piece or not. But privates can be fickle. They can ask for one thing and fall in love with another, or decide not to buy at all. And because their eyes have not been trained in the intricacies of the stone, they ask a lot of questions.

"Is this a cushion?" the young man asked my father at one point.

My father said that it was not a cushion, that cushions were his specialty. He also threw in that cushions were rounded at the corners and got up to retrieve a cushion or two from his safe, just in case the private was interested.

He didn't go all out. He didn't tell the private that original antique cushions have smaller top facets than other stones, or that they call it a cushion because the diamond's shape actually resembles a pillow. Usually, if the opportunity arises, my father likes to educate people. But that day he let the private do most of the talking. The private said his wife would come by and have a look, that she would be in touch. But this meant nothing. The private and his wife might decide to buy elsewhere. Or suddenly realize that they could use the money to take a trip to Asia or donate it to charity. Or they could buy the ring but take weeks to do so, weeks in which my father might have sold ten such rings to other dealers.

But if this sale went through, it would be different from the transactions my father made every day, transactions that often left jewelry floating back and forth between the hands of dealers. What this private had to offer was that whatever he bought—if he bought—would be worn on the body of a woman.

Chapter 5

Hashem's Diamonds

On Forty-seventh Street, it is sometimes easy to forget that diamonds are for the ears and necks of women rather than the palms and pockets of men. Yet it is the language of women that the dealers use to describe their merchandise—sexy stone, wide girdle, small cleavage—words that the ultra-Orthodox Jews of Forty-seventh Street would never use in public with regard to an actual woman.

I see them every time I'm in the district. The outer layers of their outfits are black and the inner, white. Black hats, black shoes, and long black coats. White shirts, white socks, and white tallis cloths, whose fringes sometimes peek out from beneath their jackets. Their sideburns are long, uncut. They walk quickly along the crowded street, as though they exist on a separate plane of reality, away from Manhattan's busy throngs, stopping only to chat about a diamond or to glance into the window of an exchange. Some walk with their heads down and eyes averted, to keep from looking at women who might be scantily clothed, their hands behind their backs so that they won't brush against one. They are *observant of touch,* guarding their flesh from contact with a member of the opposite sex. Once, at a jewelry show, a pin of my father's caught the eye of a Hasidic woman. She was there with her husband. My father wanted to give her the pin so she could try it on, but she wouldn't take it. At least not directly from him. He had to hand it to her husband, who passed it on to his wife.

When I was growing up, it pleased me that my father worked among so many religious people. Though I myself had never considered becoming a Hasid, I attended a modern Orthodox Jewish school, where I was taught that it was sinful to eat non-kosher food, or board a car or put pen to paper on the Sabbath. My family was different from my classmates', and when my mother drove past the street on which the school synagogue was located on Saturdays, I ducked down in the backseat of the car, praying I would not be discovered. The fact that my father worked in the diamond district with all of the black hatters seemed to redeem us a little, as though we could be made holy by osmosis. Years later I am a different person. I stopped caring long ago which Jewish laws my family does or does not abide by, but the Hasidim are still there on Forty-seventh Street, a satisfying constant. To me, the black hats and jackets are as much a part of my father's workplace as the diamonds are, as though this elaborate attire were not only a part of religion but also a method of carrying stones, providing layer upon layer of concealment.

Every weekday morning, groups of religious Jews travel from their communities in Brooklyn's Borough Park and Monsey in Rockland County, New York, into the diamond district on special buses. The earlier ones have prayer services en route to Manhattan. On Mondays and Thursdays, they read from the Torah, which resides in a small wooden box that rests on its side on the luggage rack above the seats. If a woman is on board, they hang a cloth on a hook from the ceiling of the bus to divide the cabin in two, for modesty—women up front, men in the back. I took the bus to Monsey once, to see what it would be like. But riding from the district to Forty-second Street, the bus struggling through the din of Midtown, passing Bryant Park and the New York Public Library, and continuing on into the suburbs, I realized it was just a regular bus. The proportion of black-hatted men to non-hatted men was not even that great. I was a bit disappointed.

Efraim Reiss is Hasidic, but he doesn't take the bus. His workday begins when his father, Jack, the diamond factory owner, gives him a call.

"Pick me up," says Jack.

"I pick up my father, we commute every day together. He always tells me, 'Please come at ten o'clock.' He's never ready before ten thirty."

Efraim's day starts long before ten thirty. At seven thirty in the morning, he attends services at a local synagogue. They run about half an hour on Tuesdays, Wednesdays, and Fridays, and forty-five minutes on Mondays and Thursdays, because of Torah reading.

When Jack and Efraim Reiss finally get on the road, they hold a small conference in their car. They talk about what they've sold, the deals they're in the process of making, and which merchandise they should bring to the next trade show. "We argue a lot."

They dispute the importance of the upcoming show in Basel, Switzerland. "So my father actually really wants to go, but he says that economically, maybe it's gonna be very slow, and it's a very expensive show. It's fifty thousand dollars or something. But I still consider that very important, because we had [a famous jeweler] coming to our booth and very big buyers, very important buyers."

Shows can be hard for religious dealers because they usually include a Saturday. "Meaning he has to be away from his family for Shabbos," Miri, Efraim's wife, translated for me. But, she told me, "You definitely don't lose money for keeping the Shabbos. That's for sure." Miri believes that for any profit her husband loses from observing the Sabbath, God compensates them in another way. Efraim and Miri's home in Borough Park is a lavish affair with marble floors and gilded doorknobs. The only part of the Reiss home that is not bejeweled is a small square of wall in their foyer about the size of an air-conditioning unit. They left it unpainted to remind themselves of the destruction of the Second Temple.

I'd taken the D train down to Brooklyn to visit them, pulling the longest skirt I owned that wasn't tight over my black pants when I reached the West Fourth Street stop. Visiting Borough Park is like visiting another country. Buildings have Hebrew writing on them. A mezuzah hung on the doorpost of a camera store where I stopped to ask directions. A loudspeaker on the street called out something in Yiddish. From the open door of a matzoh bakery I saw a group of men and women rolling dough; Passover was approaching.

There aren't as many synagogues on Forty-seventh Street as there are in Borough Park, but wherever he's dealing, Efraim is never far from a place to pray. When it's time for *mincha,* he can choose from a minyan at the Diamond Dealers Club, a small synagogue in one of the diamond buildings down the block, or his family's cutting factory. The Reiss staff is large enough to include the ten Jewish men needed for communal prayer. At three thirty each day, they gather in the small lunchroom between the cutting factory and the offices, face east toward Jerusalem, and pray.

Efraim and his family belong to the Bobov sect of Hasidism. On Forty-seventh Street there are various sects, with different beliefs. They start the Sabbath at different times each week and belong to different communities with different rebbes. For some Hasidim, the State of Israel is home, and for others, such as the Satmar sect, Israel is forbidden to all Jews until the apocalypse, a highly anticipated event.

Despite all the praying, Efraim considers himself a relatively lax Hasid. The only times he has ever felt a pang of unease are when he's had to decline a handshake from a woman. He is not looking to live a life of isolation from the secular world. He made sure to tell me that he also deals with Arabs, as do other Hasidim. In the district, it's all about the diamonds. After all, even the most pious of Jews on Forty-seventh Street benefit directly or indirectly from pre-Christmas business. Hashem is their only God, but each year, Jesus brings home the bacon.

The Hasidim are a little bit like the diamond district itself—the old world living inside the new. Instead of drinking happy-hour cocktails, Forty-seventh Street dealers pray together. Their corporate events are bar mitzvahs and circumcision ceremonies, and rather than holding swanky charity balls, men go from office to office, ringing doorbells, collecting Jewish alms. Old-fashioned arbitrators play the part of lawyers for diamond dealers. In the event of a disagreement, it is they who get called upon to intervene, a practice the Talmud advises. The district is filled with connections to a Jewish past, just like the diamond business itself.

The Jewish bond with the diamond trade was forged centuries ago by a history of hiding, persecution, and desperate getaways. By the time the Second Temple was destroyed in 70 CE, Jews were known for their skills in shaping precious metals, but it was the medieval times in which their affiliation with gemstones began. During the Dark Ages, most of the mercantile world was closed to Jews. Almost all of the guilds that dominated business and craft excluded them. The Jewelers' Guild was an exception. So Jews learned to cut and polish gems.

Jews were not allowed to own land, so instead, they put their funds into diamonds—the most precious of gems. These small packets of capital were then hidden and carried when pogroms broke out, or when countries declared expulsions. Jews needed to leave quickly and pack light. While their old national coinage became useless, diamonds were still precious in any country.

Diamond cities often arose in places where Jews found livable conditions. Merchants and manufacturers populated towns perched on the Indian Ocean that served as conduits between India's diamonds and the Western world. Farther west, in the late fifteenth century, the Inquisition forced Jews to abandon Spain and Portugal and move, in large part, to Holland. Their diamonds went with them. Throughout the sixteenth and seventeenth centuries, Jews brought diamonds from India to Amsterdam. These rough gems needed to be cut, so eventually Jews turned to Christian diamond polishers, who had come from Antwerp, to learn the art. Amsterdam became an important cutting city, and by the 1800s, the majority of its diamond people were Jewish. But when South Africa's diamond rush flooded its market with stones, it created a slump that drove many of the city's diamond workers back to Antwerp, where, into the twentieth century, Jewish diamond people emigrated from the brutal regimes of eastern Europe.

Over the ages, as Jews were linked with the diamond trade, Judaism became a worldwide business system in addition to a religion. Rabbis were not just spiritual guides but international deal enforcers. A gentile diamond dealer could run away to Antwerp from his obligations in Amsterdam, but a Jew would be met by a rabbi who had been in contact with the rabbi in Amsterdam regarding the transaction he had failed to honor. And just as the Jewish community became a means of regulating Jewish diamond dealings, the business adopted a religious ethical standard. The Yiddish phrase *Mazal und brucha* comes from the famous medieval Torah commentator Maimonides, who told his gem-dealing brother to seal deals with the Hebrew version of those words: *Mazal u bracha.*

The precautions Jews had to take during centuries of global persecution also set the foundation for many of today's diamond customs. My own father's secrecy comes from a long tradition of tight-lipped diamond dealers. For centuries, there was no choice but to be discreet. During the Inquisition and the pogroms, Jews kept their diamonds hidden in order to assure holding on to them. The diamond business was virtually a paperless world because written contracts were too dangerous. A man's promise was safer than his signature, and trust is still the most vital component of the trade. After all, a dealer has only to lick and sign a cachet envelope with a diamond inside, and with this gesture, he has bound its owner in fidelity.

Even bargaining is an old Jewish endowment, practiced by the Torah's superheroes. Abraham and Moses negotiated with God. Spare Sodom and Gomorrah, don't kill the Jewish people. Sometimes they won and sometimes they lost—that's business. If it's all right to haggle with God, you can haggle with anyone.

But I believe that the diamond is suited to Jews in an even more primal way, that it was inevitable for a people who have faced obliteration so many times to make the most resilient material on earth their trademark commodity. Like diamonds, Jews can be brutalized, cut down, layer by layer, and still they survive.

When World War Two came, the Holocaust purged Europe of most of its Jewish population, effectively killing off the continent's diamond trade, as much as ninety percent of which was Jewish. During the forties, those who were not murdered, imprisoned, or hidden fled, leaving the European diamond hubs to fade away.

The ones who managed to escape from Europe did not leave empty-handed. Stuffed in bags, stitched into clothing, and rolled in tissue paper were diamonds. Other stones remained on the Continent but were concealed during the war, while their owners went into hiding themselves. When Fannie Bienstock was arrested by the Nazis in Belgium, she packed a collection of small diamonds into the heel of a shoe. Her sister, who had been arrested along with her, was later released because of her husband's Iranian citizenship, so Fannie gave her the shoe. Her sister then gave the shoe to a sympathetic gentile, who buried it in his backyard. Fannie perished in the camps, but some months after the war, her son, Herman, made his way from the concentration camp of Mauthausen in Austria to Antwerp, to look for remaining family.

I met Herman through another dealer. I was at an exchange on Forty-seventh Street, at the booth of a man named Lee Runsdorf, who had been on the street since the sixties and who donned an expression of disbelief or miscomprehension almost every time I asked him a question. At first, I felt defeated. Later, I realized this was because of his bad hearing. I had asked Lee if he remembered an influx of refugees when he came to Forty-seventh Street.

He yelled over to a man, "Herman!" then, more quietly, said to me, "He loves to talk. You're in business." Then louder, "Herman Bienstock!"

A small man in a flat hat came over to Lee's luxurious booth. He had more hair than some people possess in youth. His grassy eyebrows were perched above large glasses that magnified a pair of beautiful brown eyes. His face was almost square-shaped, his mouth a horizontal line.

"Six foot eight tall," he joked as he made his way into the booth. I recognized him from the grandfathers of my youth, the grandfathers of my day school friends, who had survived concentration camps and ghettos only to be molded into smiling souls with a universal accent.

When Herman arrived in Antwerp after the war, the gentile who had hidden his mother's diamonds found him and gave him the shoe. A few years later, he wrapped the diamonds in parcel paper, put them in his pocket, and sailed to America. Herman told me that he never sold his mother's set of loose diamonds. They reside in a safe, the location of which he won't give away.

By the end of 1940, fifteen hundred dealers had immigrated to New York. That year, the city's diamond trade amounted to $35 million, an approximate $6 million increase from the previous year. "This year," proclaimed a 1941 *Harper's* article about the diamond industry, "is expected to break all records."

The Diamond Dealers Club on Forty-seventh Street was born out of the atrocities of the twentieth century. At first, most of the exiles worked in the old diamond district, on the Bowery in downtown Manhattan. The original Club stood on Nassau Street, as did its successor, an antiquated building with an elevator that members had to pull up and down with their hands. But soon, with all the new European members, the Club outgrew its headquarters. So the administration purchased a workroom from a dress manufacturer on Forty-seventh Street. They bought long tables for trading and hired a crane to install taller windows so that the diamond men could look at their stones in daylight. In 1941, the Diamond Dealers Club moved into its new home at 36 West Forty-seventh Street. Much of the district's buildings were still inhabited by vendors of ladies' garments.

As diamond men from Belgium flowed into New York, they, too, chose Forty-seventh Street. Many of the Jewish Belgian immigrants were forming communities uptown, so working on Forty-seventh Street was an easier commute. The dealers who opted for the new diamond district did well. Diamond cutting in New York also mushroomed, because it replaced Europe's cutting hubs, which had shut down during the war. A seasoned cutter could make $150 or $200 a week, a killing at the time.

After the war, the diamond district experienced a slowdown as Europe's cutting hubs reclaimed some of their former business. But over the years, New York's importance in the manufacturing world grew. By the 1960s, it was so central to the trade that many of the Bowery's most enthusiastic tenants were packing their merchandise and moving north. The Club at 36 West saw the American diamond trade through its most prosperous times.

"It was like the New York Stock Exchange," David Abraham, the former VP of the Club, told me—if the New York Stock Exchange had a Jewish restaurant in the middle of the room. The floors were linoleum and the color scheme unremarkable, so as not to distract from the important colors: D, E, F, G, H, I.

The Club at 36 West was a loud and frantic place, made more chaotic by the absence of technology. Dealers didn't have cell phones on which to carry out international negotiations. When someone called the Club for a member, he was put through to one of several female switchboard operators. The operators then summoned the member through the intercom system, assigning them a phone booth number. Often there was so much foot traffic on the floor that by the time the man reached the telephone, his caller had hung up. Deals occurred at the speed of light. Diamonds changed hands several times a day. Foreign dealers stopped by to purchase and get a sense of the American market. You could hear six or seven languages on the premises, most commonly Yiddish, English, and Hebrew, but also French, Spanish, and Persian.

When I was talking to David Abraham in the Club café, he asked a diamond broker with an Israeli accent who had happened upon our interview, "It was nice there, no? The business was nice there."

"Beautiful," said the man, as though he were admiring a work of art.

For a few minutes the two reminisced, finishing each other's sentences. "Tons of business, tons of business," said the man nostalgically.

"There was a lot of good business. Used to be a lot of big stones here," David said. "I used to have people waiting on line in my office come up to see it. They used to wait on line, sometimes for an hour, hour and a half, two hours."

But it wasn't just the diamonds that made the Club. Most members had either survived or fled the greatest tragedy of the twentieth century. If they were not survivors themselves, they knew survivors, or victims. A great many of today's Forty-seventh Street dealers are no strangers to disaster. Wartime Europe, the revolution in Iran, the Soviet Union—this shared history of tragedy is an undercurrent that binds the neighborhood together and makes its members protective of one another. After all these centuries, it is still runaways who are drawn to the diamond business—my own grandfather deserted the Russian army in the aftermath of the Second World War. But on Forty-seventh Street, the diamond dealers found a community of people whom they could trust with the most valuable of gems, and with their own safety. Which is why what took place in 1977 was the worst thing that could happen to the men and women of Forty-seventh Street.

My father was still in Germany when the district experienced its first diamond murder. It was Tuesday, September 20, 1977, the day before the eve of Yom Kippur, the Jewish holiday of atonement. Pinchos Jaroslawicz, a twenty-five-year-old Hasidic broker, was over an hour late for dinner. Some wives wouldn't think twice about the delay, but Pinky was never this late. His wife called her father-in-law, and with a group of friends, they went to the police station. A close family member of Pinchos's who did not want to be named told me that the police refused to file a missing persons report, because it hadn't been twenty-four hours since Pinchos had last been seen.

So at ten p.m., Pinchos's father and wife went to Forty-seventh Street themselves. They searched those buildings that hadn't yet been locked up. They asked anyone they came across if they had seen or heard from Pinchos.

Pinchos's family returned to the diamond district the following day, but they didn't find anything. Wednesday night was Yom Kippur. Forty-seventh Street was empty; everyone was in synagogue. The search was put on hold.

Slightly incredulous, I asked Pinchos's relative whether anything was done on Yom Kippur itself.

"Kippur itself?" he said. "Yes. We prayed. All over."

Pinchos was known as something of a saint. People entrusted their diamonds to him without asking for a memo receipt. A week before he died, he wrote a holiday check to an impoverished man in the neighborhood where he lived. Mistaking the amount on the check for a hundred dollars, the man tried to use it at a grocery store. When the grocer refused to cash the check, saying it was too much to make change for, the man looked at it again and saw that Pinchos had given him a thousand dollars.

On the Sunday after Yom Kippur, roughly a hundred people showed up in the diamond district to look for Pinchos Jaroslawicz. But they didn't find him. Three days later, his body was discovered inside the office of a diamond cutter in a plastic bag beneath the tool bench in his workroom. The cutter claimed that two masked intruders had broken into his office, robbed and killed Pinchos, and forced him to bind the body.

The cutter and his accomplice, also a cutter, were charged with murder. Hasidim filled the courtroom in which Pinchos

Jaroslawicz's murderers were tried. They wanted to show people that Pinchos was an honorable man, part of a community. The jury found the cutter and his accomplice guilty; they were given twenty-five years to life.

Pinchos's murder stunned the district. Although Forty-seventh Street lay in the center of one of the biggest cities in the world, the crime was like a small-town homicide. The sin had come from within. Both killers had worked in the district.

"A murder?" one dealer said to me, remembering the shock. "We never had a murder before. That's crazy. We had robberies, we had thefts, we had holdups. Murders? That didn't happen."

Had the diamond people come all this way, from the four corners of the earth—places where they had been hunted like animals—had they suffered all those unspeakable things to see one Jew kill another for some stones?

The Yom Kippur Murder, as it came to be called, changed the very nature of diamond dealing. Many of the stones Pinchos had been holding disappeared when he was killed, and because some of their owners had not taken a memo form, they were unable to declare the losses with their insurance companies. According to Pinchos's relative, diamond men became more leery of handing goods out on consignment without taking receipts. Forty-seventh Street shops started opening later and closing earlier; no one wanted to be the first or last on the street.

As *New York Times* reporter Murray Schumach pointed out, Pinchos Jaroslawicz's murder also changed the relationship between New York's diamond community and the city's police force, which had been strained for years. In the 1970s, Forty-seventh Street dealers were notorious for refusing to talk to the police. But after Pinchos disappeared, the Diamond Dealers Club leadership told its members to talk, and they did. In a way, the killing collapsed the borders that separated the Forty-seventh Street enclave from the rest of the world. The diamond people had been awakened.

Chapter 6

The List War

The diamond district of the seventies remained a labyrinth of secrets. You could enter as a customer but never as an insider. Even diamond prices were an internal matter, only for the eyes and ears of the diamond dealers, their diamond street, and their Diamond Dealers Club. And it was during this era that a man named Martin Rapaport—a fellow Club member! a fellow Jew!—revolted.

I met Martin in the late spring of 2008 in Las Vegas, at the Venetian hotel. He is a hard man to catch—he splits his time between Vegas, Jerusalem, New York, Mumbai, and various other outposts of his empire. I waited near the man-made canal where the gondolas float in aqua-colored water. When the harried young woman coordinating Martin's schedule called to say he was ready to see me, I passed under the Sistine sky that covers the hotel's cab dock and made my way inside. The Venetian pumps gusts of fragrance through its lobby vents to mask the cigarette smoke, so the lobby smells like a woman who has just had a cigarette, then sprayed herself with perfume. The coordinator—a young, attractive Parisian—and one of Martin's sons led me through the casino and upstairs, into Martin's suite. It was as large as my entire apartment. Martin told me that he has stayed in this same room each year for ten years. "It's a tradition."

Though he may be a lonesome visionary in his ideals, he is constantly surrounded by an entourage. Rapaport employees chatted in the bedroom. People wandered in and out. One man was extemporaneously invited by Martin to a formal dinner—a space had opened up because the State Department couldn't make it. "Their legal department won't let them come to the dinner," he explained.

While we spoke, Martin sat in an armchair with his back to the lights of Las Vegas. I sat on a couch next to the armchair. Another one of his sons was stationed at his other side. Martin has ten children, and three of them accompanied him on this trip. Despite its size, the Rapaport enterprise is a family affair.

Martin is a round man—round body, round face, round eyes. He wears his dust-colored beard and mustache fluffy but not long. He sports a round skullcap, because he is a modern Orthodox Jew. Usually, his voice is thunderously energetic, but by the time we met, it had been a long day for him. He had already changed into plastic Crocs, and sometimes closed his eyes, whether in concentration or fatigue, I was not sure.

Before we got started, he wanted to know what I was writing about. Then he interrupted to try to figure out which other Oltuskis he knew. Then Martin asked me what my book had to do with him.

I said that, as he might know, he was fairly important in the industry.

This is an understatement. Almost every reputable diamond office subscribes to Martin's newsletter and magazine, the *Rapaport Diamond Report,* which covers worldwide trends in diamonds and jewelry. Martin's online trading service, RapNet, is the largest of its kind, posting billions of dollars' worth of stones. His company, the Rapaport Group, is an international corporate kingdom that transports diamonds, grades diamonds, does everything except own diamonds. But most of all, Martin is the man who started the Rapaport Price List, the piece of paper that some people believe dictates the cost of diamonds.

"Ehhhh," he whined modestly at my compliment. "The guy everybody loves to hate. Okay, so—" He interrupted himself with a laugh. "That's not true. Most people like me." There was a brief interruption, while Martin dramatically stopped hotel staff from taking his refrigerator (they thought it was broken and had come to replace it), and then he got back to our interview.

"So what do you wanna know about Rapaport?"

Martin began his diamond career in Antwerp as a cleaver, splitting rough diamonds in two before they were polished. He went on to work for a renowned dealer named Feivel Doppelt. There, he sat at a corner of his boss's enormous desk and sorted diamonds by customer. Because of its more commercial market, America was allotted the lower-end diamonds, whereas Japan, whose consumers were willing to pay a premium, received the more expensive stones. Martin learned that the essence of the diamond business lay in maintaining an equilibrium between what each customer wanted and what sorts of diamonds a company had available to distribute.

After Feivel Doppelt, he moved on to the merchant side of the business. He came to the United States and started brokering diamonds by knocking on the doors of various jewelry shops on the west coast of Florida. Martin didn't want to be like other diamond people. He didn't even want to own diamonds. Instead, he wanted to connect those who did. He was in love with data, a nerd more than a merchant, but still, a businessman through and through. Diamond prices were what he found interesting, and the seventies were the time to pay attention to them.

The economic world was balancing on one foot. Inflation was rampant, oil costs high, and interest rates terrible. Because the U.S. dollar was soft, Americans were investing in more tangible items. They bought art. They bought precious metals. And they bought diamonds. The move toward speculation in precious objects occurs every once in a while when the economy slumps. Like investors, dealers also buy diamonds for resale value. But the success of the diamond industry is predicated upon the fact that customers want diamonds for love and vanity, not speculation. As laymen bought investment diamonds, prices rocketed dangerously.

This was the age of third parties, Martin said. The GIA was founded in 1931 and had been grading diamonds since 1955, but as the dollar became less reliable and more individuals began investing in diamonds, more stones were sent to the labs to be given a grading report. After all, an item of investment needs to be evaluated objectively. During the late seventies, diamonds came closer to being a measurable commodity than ever before. In order for an item to be quantifiable it needs two components: a standard way of classifying its quality (from labs like the GIA) and a matching price for those classifications. Martin would supply the latter.

As a broker rather than a diamond buyer or seller, he was a third party. He saw both sides of transactions and knew how much both parties were willing to pay and receive. During his career, people constantly approached him for diamond prices.

In July of 1978, he answered in a big way. He released a list of "high cash New York asking prices," wholesale prices for polished diamonds. The list came out each week and stated, in table form, the asking price per carat of both round and pear-shaped diamonds, by color and clarity. Martin called it the *Rapaport Diamond Report* and sold it to anyone who was willing to pay twenty-five cents per list. It was available not only to members of the trade but to the rest of the world, as well.

He describes it as the first of its kind. There were other lists, but they were published either by outside investors or by insiders who owned diamonds themselves. Martin decided that he would not own diamonds, so that he could be objective. He believes that this is why his list succeeded. I think that his loud voice may also have had something to do with it.

The list was a way to differentiate himself from other dealers and brokers. "If your business is just to control or to have inventory, why would you publish a list? Why would you want anyone else to know what the prices were? See, you had to have a totally different mind-set." And Martin did. "You gotta understand, it's kind of like, you change the world by publishing information. You're giving people information. You're democratizing things."

Martin was also helping private customers by allowing them a glimpse into how much their diamonds cost at wholesale prices. He was a democrat, a crusader for transparency—one of his favorite words—in one of the most opaque businesses in the world. He took the mysticism out of diamond dealing. All of a sudden, people who knew nothing about diamonds, who hadn't spent their lives learning the intricacies of the craft, could consult a chart and find a stone's wholesale value. For so long, the price of diamonds was merely what a customer was willing to pay. Now customers no longer needed to take a dealer's word for granted.

The way people looked at diamonds would change, too—from an obscure science understood by only a few, to a quantifiable product like any other. According to his list, two diamonds of the same weight, shape, and clarity should cost the same.

The list didn't speak of those airier qualities. The way one stone shines more brilliantly than another, despite their identical clarities and colors. The fact that diamonds originating from certain mines shimmer vividly, whereas others glare starkly. That some diamonds can be called beautiful, whereas others, of no lesser gradable quality, cannot. It said nothing of a stone's life.

There were implications for dealer-to-dealer trading, too. Before, the cost of a diamond was more subjective. It was up to sellers to rate the stones and buyers to accept or decline values. Along with the GIA and later the Internet, the price list transformed diamond dealing from a business of individual scrutiny and personal opinions to a business of global terms.

"You became obsolete," Nat Trauring, who has been on the street for over sixty years, told me. With the dominance of GIA certificates and the Rapaport list, "you're not an expert anymore."

Not everyone thinks that diamonds should be a business of universals. Sal Lipiner, who was on the board of the Diamond Dealers Club when the Rapaport Price List first came out, told me, "I remember debating Martin, and I said to him, 'Have you ever seen a woman wear pork bellies around her neck, or soybeans? You can't commoditize this.'" Sal doesn't think "a woman receives a ring and gets all excited by it because it has a nice lab report. I think it's because of what the ring represents: love, magic, whatever you want to call it. I'm still in that era."

When my father was just starting out in the business during the late seventies, he witnessed dealers arguing with a young Rapaport in the trading room of the Diamond Dealers Club. One or two shouted at Martin. Then a third joined in. Martin kept up, defending his price list. My father found the spectacle of their loud quarrel somewhat comical. But he, too, thought Martin brazen for telling others how much their stones should cost. Like most, he believed the Rapaport Price List would be short-lived. Like most, he was wrong.

"We were very upset," said a man who manufactured diamonds on Forty-seventh Street at the time. "We felt it was very wrong of him to do it." Gold and silver double or triple in price once they've been transformed from bullion into finished jewelry. But Martin wasn't listing the price of rough diamonds. "He was listing the price of the finished goods, so how in the world were we supposed to make a living? So it was very hard."

Yet despite people's objections to the list, there was an underground force that simultaneously drove its popularity: curiosity. Even those dealers who opposed it desperately wanted to catch a glimpse of its figures. I learned from a former employee of Martin's that instead of buying the list, dealers would photocopy it from those who had paid for it.

In so doing, they actually helped circulate the list, to make it a staple of the industry. A few years into publication, Martin would start printing on red paper, which produced a hazier duplicate and was harder to xerox.

The first two years of the list were relatively calm. Because of the investment craze, prices were strong, so Martin had good things to report. A round brilliant one-carat D flawless, the industry standard, increased from $22,000 in July of 1979 to $39,000 about four months later. That same diamond cost $63,000 in February 1980.

Things couldn't stay that way for long, though. Speculation and investment diamonds had pushed prices to unreal heights. Interest rates shot up, and the diamond investment bubble started to burst. Prices fell quickly. By the end of May 1980, the same one-carat diamond cost $52,000. For a few months there was a lull. Then prices dropped again in 1981 to $41,000. By July 1982, that D flawless cost only $18,000. It was a disaster.

When prices plummeted, the other list publishers drew back, but Martin and his people kept printing. "Think about it," David Abraham told me. He was at the Club during the whole affair. "People had bought diamonds on borrowed money. The interest rate they had paid on these loans to buy the diamonds had suddenly gone up from like seven, eight percent to twenty percent, twenty-one percent. They couldn't sell." Still, David felt that Rapaport was not to blame: "You can't stop a person from doing what they think is right, even if you disagree with them." But back then, many dealers disagreed with David.

It was not just Forty-seventh Street that was displeased with the list. From Antwerp to Tel Aviv, diamond people voiced their condemnations. In 1982, in an effort to restore the old order, the World Federation of Diamond Bourses disallowed price lists. But Martin kept publishing. He wasn't going to deny the public information. He felt the price list was most vital during this unpredictable economy. Which is why he kept publishing even after several rabbis threatened to take him to the Beit Din, the Jewish court. One man was so enraged that he chased Martin around the Club at 36 West. "A lot of people were yelling at me that day, 'cause we had dropped the price list a lot." The man didn't get to Martin. "I just kept running."

Then one Thursday night, on the eve of releasing his latest price list, Martin received a phone call from a man who implied Martin might soon meet his death. The threats kept coming, usually on Thursday nights. "We're gonna kill you," people said. "You should die." Once they even sent an ambulance to his New York office.

Martin got himself a bulletproof vest. He wore the vest everywhere, even to his graduate economics classes at NYU. But he kept publishing. He had the support of his wife, who felt that what he was doing was important.

People blamed him for the crash, for their hardship. "I became kind of like a scapegoat," Martin told the press.

He pitied those who had lost money with the crash, the crash whose numbers he detailed in his list. He himself knew what it was like to go bust. He'd once lost all his money working as a sugar dealer in Israel. Still, he believed in giving everyone the same information. No tricks, no secrets.

The showdown intensified in the summer of 1982, when Martin was cited in the magazine *Collector Investor* as saying about the industry: "Morals, ethics, feh! If the devil himself showed up they would sell to him." In his suite in Las Vegas, Martin spelled out the word *"feh"* for me. "And they actually threw me out of the New York Diamond Dealers Club citing that quote." They gave him a yearlong suspension.

The point Martin was making in the magazine was that, while the diamond people abided by a remarkable code of ethics within the business, they were not as concerned about what happened once the diamonds left the industry and emerged in the outside world. People were selling investment diamonds to others, who were getting burned.

Sal Lipiner, the former board member, tells a different story about Martin's suspension, one that has nothing to do with magazine quotes. "Because of this question of where do these numbers come from or how does he determine these numbers, the board of directors said, 'You can put out a price list. You can put out any kind of price list you want. Any member can put out any price list they want, as long as they possess the merchandise that's on the list and are willing to sell it for the price that's on the list. You can't invent a fictitious price.'" Lipiner insists that Martin was expelled because he published a list of prices for stones he didn't own or sell.

That year, Martin sued the Club for kicking him out, because without the Club, he couldn't get his numbers, and without the numbers, he had no price list. It was the first time anyone had sued the Club on grounds that were not related to an unfavorable verdict in the Club's arbitration system. The courts ruled in Martin's favor, and the following year, he was back in the Club with a court order from the New York State Supreme Court in his hand.

But at the Club, Martin was bullied. Someone even called him Hitler.

In 1984, he told the *New York Times*, "Forty-seventh Street has to recognize that it's in America." But this was the essence of the price list war: Forty-seventh Street *wasn't* in America.

It was a vestige of the older world, a street for diamonds, not First Amendments. So was Tel Aviv. So was Antwerp. Martin Rapaport had violated the sacred law of secrecy. Throughout the diamond world—filled with stockpiles, closet prices, stones in chest packs and pockets and valises—everything hinged on the ability to keep the most valuable things hidden.

Despite the tension, Martin was not the type to keep a low profile. After he reentered the Club, he brashly ran for its board, but "diamond prices had plummeted just before, so *that* wasn't very good for my election," he said to me in Vegas, smiling.

Eventually, people learned to live with Martin. In 1987, he became a board member of the New York Diamond Dealers Club, just five years after they'd suspended him. Martin attributes the transformation to a hunger for change, for transparency, which happened to be his campaign platform. (The year he won, the Club published its budget for the first time ever.) It also didn't hurt that he had gone from office to office collecting signatures to get his name on the ballot, getting to know his future constituents. Some people told him to go away, but others appreciated the initiative.

Though resistance toward Martin did not stay as intense, it didn't just melt away, either. "It's like a husband and wife," said Eli Haas, who dealt at the Club. "You know, sometimes it's okay, and sometimes they have a fight. The next day, they're going to the movies together and everybody's happy."

Every so often, Martin's list elicits almost as much controversy now as it did during its early days, especially when diamond prices get too high or too low. Martin claims that the list is just his opinion, a reflection of market prices and a basis for negotiation. But many dealers believe it actually influences prices. "It's very scary to have one person that controls this, that has so much power by controlling the list. No watchdog, no nothing," an anonymous diamond dealer told me. If Martin reports a price increase in three- to four-carat diamonds with very small inclusions, it will be hard for buyers to bargain down when purchasing three- to four-carat VSI diamonds that week. By releasing a directory that provides both dealers and customers with a concrete number, Martin is often held responsible for tight profit margins.

David Abraham believes Martin just reports what he observes in the market, but others see it differently. Mr. Lipiner, for one, is convinced the list affects prices. "If for no other reason than, despite all protestation, the ultimate consumer gets his hands on that price list just as easily as anyone else does. Now, if the ultimate consumer walks into a retail outlet and says, 'Hey, I've got a price list here that says I could buy this item for a thousand dollars,' the retailer's kind of hard put to say to him, 'No, no, no, no, it costs twelve hundred.' Customer will say, 'No, no, see here. It says a thousand.' Now the retailer's going to have to buy it for less than a thousand. And now we've created something that has a fictitious price that John Q. Public walks around with, that forces an industry to somehow function around this price that nobody knows the origin of."

"But doesn't that almost make it real, if everyone is forced to function around it?" I asked.

"Well, it doesn't, because nobody pays it. You're paying over it or under it."

If you tell Mr. Lipiner that the list has created a range of prices for which a particular diamond will sell in a given week, he will say to you, "There always *was* a range. So if you were a little bit of a sharper buyer, you bought it for two percent *less,* and if you were a little bit less sharp, you paid two percent *more,* which is the same thing that happens now." But when Martin reports a drastic change in prices, everyone listens. Whether you believe that the Rapaport list influences the market or the market influences the Rapaport list—and most likely, it is a combination of both—the price that dealers will pay for a certain type of diamond is now intimately connected with Martin.

One thing Martin and Mr. Lipiner agree on is that haggling is just as complicated a game as it was before the list. Only now dealers haggle for percentages above and below the list. "Twenty below." "Fifteen above." A chorus of discounts and premiums on Rapaport's list rings out daily in the offices and exchanges of Forty-seventh Street, and everyone, even Martin's critics, has no choice but to join in.

But perhaps the most controversial aspect of Martin's list is that the method by which he determines his prices is still a mystery; it's the one thing he is not transparent about. "Sometimes," my father told me over the phone one day, "you almost have the feeling that he looks into a crystal ball, you know?"

To this day, Martin believes that the misfortune he suffered as a result of his list only added to its validity, because "I was willing to suffer bomb threats and tell it like it is, 'cause we drop prices like crazy when we have to."

His struggle also had repercussions for the Club. Lawyers, who were an anomaly to the Club's arbitration process pre-Rapaport, became more common. The Club was also more hesitant to suspend a member after Martin sued them and won. "I think people weren't as afraid of the Club as they used to be. I had broken that ox. I'd shown that, if there's injustice, you can stand up against it."

Most of all, the dominance of the list revolutionized the way diamonds were traded. On Forty-seventh Street and in many districts around the world, the colorful two-page weekly list, informally called a Rap Sheet, is omnipresent, xeroxed and tucked into wallets and pockets, posted on the walls of offices.

But not every stone has a place on the list. Aside from some conversion tables and indices, it prices only certain shapes, sizes, and colors of diamonds. In this way, the list benefits dealers of unusually shaped, sized, or colored stones, for in creating a region of classification, it has also created a region beyond classification, in which dealers determine prices according to their own judgments.

My father's newest niche, Cape stones, are technically below the priceable regions of color. Their complexions—anything from M to Z—run from mildly yellow to the color of watered-down apple juice.

"Can you compare the color to something?" I asked my father.

"I don't know, maybe a cup of tea," he said over the phone, but our connection was bad, so I thought he said "pee."

"Oh!" he wrote to me in a text message. "That would be more like a fancy yellow."

Fancy yellows, anything below Z, are diamonds with such rich yellow coloring that they are considered highly desirable. Because Cape stones don't draw the same kind of premiums that whites or fancies do, my father can afford to buy them larger. When he trades in them, there are no lists to consult, so he and the dealer on the other end of the transaction make up their own prices, bargain the way dealers used to bargain before Rap Sheets. Which is why when he and Roy bought their fifteen-carat Cape, my father simply called up his friend and said, "You know that stone we bought and sold. . . . Well, try to remember it." It is why I stood in Roy's office that day and listened to him use nebulous words like "ugly." Why my father and the dealers on the phone haggled with no reference to twenty below or five above. There were no lists to rein them in.

My father likes this no-man's-land, where he is free to set prices without the world or Martin Rapaport encroaching on him. But for the most part, the list is a fact of life.

It wasn't Martin's only innovation. While his list turned the diamond world upside down, he forged ahead with other projects. As he and his group gathered prices each week, they became privy to the needs of buyers and sellers. They knew who was looking for a two-carat flawless and who had one to sell. They knew who wanted to get rid of small stones and who was hoarding them. It was the stuff of pure brokerage, but Martin wasn't going back to knocking on doors. He had his eye on something bigger: large-scale matchmaking. A diamond database.

It began with a Model I RadioShack tape recorder that had a 4K memory chip—"which is less than this watch has," Martin told me—and a decision to give the public access to even more information.

"I hired these girls who weren't diamond people. I said, 'Anytime someone calls and wants to buy this item, stick a red thumbtack down, and anytime someone wants to sell, put a green one, and whenever you see a red and green in the same box, get on the phone and do the deal.'" Eventually, the thumbtacks developed into RapNet, the online trading service.

In addition to the database and network, Martin found another market: diamond education. People had been asking about the reasons behind the increases and decreases on his list. The answers were complicated, greater than just Forty-seventh Street. So Martin started the *Rapaport Diamond Report,* to provide the bigger picture. He has become a middleman to middlemen, an intermediary between the known and unknown worlds. For customers, he demystifies prices; for dealers, he demystifies the reasoning behind price trends. His company tagline is "information that means business."

Each time Martin tries something new, he creates a new category of diamond service. He and his list are no longer rogue factions but rather mainstream sources on diamond values. Even private customers have been told that when they go to Forty-seventh Street they deserve to buy a diamond at Rapaport value, no more. So they bring along their Rap Sheets when they hunt for diamonds, and the list offers them a degree of comfort as they enter the jungle.

Chapter 7

The Older World

The true authority in my father's life isn't nearly as famous as Rapaport, nor as dramatic. He doesn't use online databases; he doesn't know how to. The newsletters he reads are written in Yiddish, and, though he visits the Diamond Dealers Club every day, he has never considered running for office. He was a merchant before he ever set foot in the Club, before he even set foot in America. He learned the rules of buying and selling in a country where free trade was considered a sin, and came out of hell with diamonds in his pockets.

My grandfather uses three names, depending on what part of the world he is in. Medicare calls him Jack. To his friends and in the old country, he is Yankel. And sometimes, if she is feeling particularly anxious, my grandmother uses his Ukrainian name, Jakob. He is a man of elegant taste, favoring gold cuff links and quality fedoras. The pockets of his best shirts are embroidered with "OJ," for Oltuski Jack. But all of these embellishments come from a different world than he did.

Most Forty-seventh Street family businesses started somewhere else, in European cities or faraway villages, generations ago. Working at the office, I began to understand that when my father bought a diamond, he was purchasing on behalf of generations of Oltuskis—Oltuskis who had lived and died—and with the predilections of one man in particular on his mind.

Opa Yankel found diamonds in the darkest place in the world: Germany after the war. My grandfather's history came to me through his stories, which, like the Torah, involved walking long distances. But rather than bringing him to the Promised Land, Opa Yankel's adventures led him to the diamond district. In the deserted stretch of earth known as Siberia, my grandfather became a businessman. It started with salt.

Yankel and his friend stood at the base of the hill of grainy white. They filled everything they could with it. Yankel poured a whole world of salt down his pants and tied the cuffs tightly around his ankles with his shoelaces. It was the middle of the war against Germany. As Russian soldiers, they had been sent by the army to work on a commune in Siberia, growing produce, since there was a shortage of farm workers. They were on their way back to Moscow's army base when they came upon the hill of salt. The train they were riding had stopped at a village near the sea. Someone must have brought the salt from the sea to the station platform and then abandoned it. So Yankel and his friend jumped off the train to grab some. No one stopped them.

They knew that they would eventually come upon a city where salt was rare and therefore highly desirable. In Eastern Europe certain natural resources existed only in specific towns. Not like in America, Yankel told me. If it's in one city, it's everywhere. Russia was different. "A couple of miles away, there would be no salt. There was something else, I don't know, maybe chicken."

When Yankel and his friend got back on the train, they poured the salt into their rucksacks, into their hats. Then they continued to journey across the vast Russian landscape. In those days, it seemed like the whole world was Russia. Even Yankel's own hometown had gone from being part of Poland to being part of Russia. After the war, it would become Ukrainian. These were fickle borders, the kind you couldn't count on, the kind that transected one place when you left and another when you tried to come back home.

On the train to Moscow, Yankel and his friend passed impoverished villages with humble houses. The townspeople would gather at the stations in the hopes of trading goods with the soldiers aboard the trains. Sure enough, one of these terminals was located in a saltless town in Kazakhstan. Waiting by its train station were the villagers, mostly women. Their men were in the army, just like Yankel. They came with glasses, which they used as measuring cups, ready to sell their meat and tobacco.

"We can trade," said Yankel. "One glass salt and a glass tobacco." So he and his friend traded their cups of salt for cups of tobacco and also took some meat. At their next destination, in a city near Moscow, they found a market and sold their tobacco for rubles. They made quite a bit of money.

"So, at once I became a millionaire." Yankel laughed. "Overnight."

Theoretically, Russia was a communist state, but as it turned out, communism had bred more than a few capitalists. In Yankel's hometown, men and women were always trading one thing or another—grain, livestock, groceries. "By us, every single person who was born became a merchant." Yankel understood that everyone desired something, that if you could provide them with what they coveted, then they might do the same for you. Trade is the same everywhere. The world is made up of buyers and sellers.

By the time Yankel jumped off that train to get the salt, he had already chosen capitalism over communism, subsistence over the Mother Country, and rubles over the Dictatorship of the Proletariat.

Three years later, Yankel was a fugitive from the Red Army. It was 1945, and he was on his way to Germany. He had fought the last year of the war as a junior lieutenant, but the army still had a larger project under way. They were building a nation. When Yankel got into a spat with an army official over an assignment, he knew he was in trouble. Russian soldiers were known to be anti-Semitic. He didn't want to risk a revenge deployment to some perilous location. So when he was granted a twenty-day leave, he took it and fled to a city called Szczecin, where he met his father and one of his brothers. His mother, two other brothers, an aunt, and several cousins had perished at the hands of Nazis. The remaining three planned to leave godforsaken Eastern Europe and meet in Germany, because Germany was where people got tickets to America.

Yankel caught a ride in the back of a military truck to Berlin. Once there, he found a displaced persons camp. The camp was filled with Jews, but no one would talk to him. "They were scared of me, saw that I was militarily dressed." They suspected he was a spy.

Finally, a man who saw what was happening approached Yankel and offered him a place to change. "He led me down to the cellar, took everything off, threw it away." During his service, Yankel had made sure not to look like a slob. He had even found a tailor in one of the cities where his unit had passed through and got his uniform taken in here, let out there. But none of this was important in Berlin. Yankel put on the civilian clothing he'd carried with him. He disposed of his big bronze medals, cut the military buttons from his jacket, and sewed on new ones. He couldn't risk being caught by the Russians, nor could he afford not to use the warm wool jacket.

A few days later, he was joined in Berlin by his father and brother. The three men made their way from the city's Soviet Zone to its British Zone, crawling over the snowy hills at night to avoid border control. From Berlin, they traveled to a small city called Zeilsheim in Germany's American Zone, and then on to another called Lampertheim. As they moved away from Berlin, Yankel was inching closer and closer both to Dora, his future wife, and to his first diamond.

Though he didn't know he would marry her, he had already met Dora, had known her since he was a boy. She lived in a nearby village, and he would come by the restaurant her parents owned on his way to other towns for business and sit with them, talking. After he left, Dora's mother would say to her daughter, "Yankel will be for you." The fact that he was her first cousin didn't matter in those days. First cousins could fall in love. But marriage wasn't on Dora's mind yet. All she knew was that when she stood on her porch watching Yankel pay a visit to other girls in her neighborhood after leaving her parents' home, she felt a surge of jealousy. Then came the war. Dora and Yankel's daydreams were replaced with strategies of staying alive and dodging Nazis. Dora spent the later war years with the Russian partisan forces in the forests of eastern Europe. Yankel hadn't seen her since.

Lampertheim was a strange place in that no one lived in his rightful home. After the war, the Americans had seized houses from German citizens—some of whom had probably seized their houses from Jews—and given them to Jewish survivors to create a DP camp. So Jews lived in German homes. They cooked in German kitchens—if they had food to cook. They walked on German rugs and sat on German chairs, and because the Americans who'd kicked them out had done it quickly, without cleaning up much, they looked at German pictures on the walls and spent their nights on the linens that Germans, possibly even Nazis, used to dream on. One of these homes was now inhabited by Dora and her roommate.

By the time Yankel arrived, Dora had received multiple marriage proposals. There was a man named Charitsky, a Lithuanian, and, she will tell you, sometimes the camp policeman. They all wanted her hand. She knew Yankel was alive—she had been in contact with his father—but she didn't know whether he would return for her. No one had thought of such details during the war. So she didn't tell Charitsky no. Nor did she tell him yes. Instead, she procrastinated. She said that she would discuss marriage when her uncle came—she had no parents to give her away anymore.

The day Yankel arrived in Lampertheim, Dora was attending a wedding with Charitsky. During the ceremony, her roommate came to tell her there were five Russian men looking for her.

When Dora heard the news of their arrival, she ran home and Charitsky chased after her and said, "Dora, *nu,* give me the word."

"No," said Dora. "I'm sorry. My groom has come."

When she saw them, she was filled with delight. She kissed her uncle and two cousins, and greeted the doctor and the lieutenant they'd picked up along the way. Since her apartment was large enough to accommodate everyone, the five men moved in, and, as Yankel puts it, "the flirtation began." Yankel would sit by Dora's bed at night.

"What did he say to you?" I asked my grandmother.

"I remember what he told me?" But later on, she did remember. He told her he loved her. And he continued sitting by her bed at night. He kissed her and hugged her, too.

After a few months of spying, Yankel's father declared, "They're already too close."

Dora's mother didn't live to see the fulfillment of her prophecy, but the nuptials took place on a winter day in Zeilsheim, at the DP camp where Dora's sister lived. Fruit, cake, and hamburgers were served at the reception. Yankel wore a suit he got from the United Nations Relief and Rehabilitation Administration, the UNRRA. Dora borrowed a veil from a German woman and a dark dress with star-prints from her married girlfriend. She hadn't the money to buy fancy slippers, but a shoemaker at the camp made her a pair of sneakers. "When I stood under the chuppah in Zeilsheim, I cried," Dora told me. But she didn't know whether she was crying over the fact that her family, who had perished during the war, was absent, or because of how much the shoes, which were unbearably small, hurt her feet.

An old black-and-white photo shows the bride and groom in a corner, near a window with a shoddy curtain of ragged lacework. A small basket leans against them. It holds a sparse bouquet of baby's breath. Yankel's head is tipped in toward Dora's. Both of them smile; her lips are parted, his sealed.

Less than a week later, they returned to Lampertheim. Every once in a while, there were problems. The Lithuanian and Charitsky tried to chase Yankel and the other four men out of the camp, claiming they suspected them of espionage. Or sometimes Dora would see Charitsky on the street and would be filled with a sense of shame, forcing her to cross to the other side. Other than that, life in the DP camp began to resemble real life.

But the camp was not only a home—it was a marketplace. After the war, the world was filled with people who had lost everything, and so the price of things was fluid. After all, trade is what a man needs or wants at a particular instant. Nothing more, nothing less. Of course, these were not legal markets. Most Jews who ended up in the camp had no money of their own, so those who traded tended to use the provisions given to them by the UNRRA.

There were no borders separating the camp from the village, so a Jew like Yankel could spot a German with a briefcase inside the camp, and ask the man, "What do you have to sell?" And on this particular day, this particular German was selling diamonds.

He was carrying six pieces of rough, uncut stones with no brilliance to them yet. They could have been mistaken for common rocks. To Yankel, they looked like glass. The only other diamond he had ever seen in his life belonged to a window maker in his hometown, who had a stone he called "dimindl" affixed to a handle, which he used to cut windows. Yankel had never actually touched a gem. He didn't know quality diamonds from window-cutting diamonds. He took one piece of the rough in his hand and examined it with his naked eye. The rough was about one carat, a fifth of a gram, nothing special in today's diamond district. And yet it is likely that over two hundred tons of earth and at least a billion years were involved in uncovering it from the ground. And when Yankel held this diamond, it was as though he held two hundred tons in his hand, as though the diamond had waited a billion years underground just for him.

Though this was a new experience, a new product, Yankel was no stranger to trade. There was the salt in Siberia, and a host of other minor goods peppered throughout his childhood. He had been buying and selling since he was a boy. Pig's hair for brushes. Chickens. There was a man in town who paid for empty vodka bottles, so Yankel collected those, too. Other times, he and his older brother would travel to Ukrainian villages with a bag. The Ukrainians had chickens, which meant that they had eggs. But they didn't eat their eggs—who could pass up such a valuable money-making opportunity? Yankel took one side of the tracks, his brother took the other, and they went door to door buying the Ukrainians' eggs, which they would resell to a more important merchant.

"Lazy, I was not."

Sometimes Yankel treated himself to an ice cream or simple clothing with the money he earned. But he also saved doggedly. He had a goal: a suit. Elegance was a priority for him even then, and he knew that a good suit was important in life. On Shabbos, local teenagers would stand around comparing their Saturday clothing, and Yankel wanted to measure up. He had his first suit made when he was eighteen years old, dark blue wool with broad stripes. It was beautiful, luxurious, much better than ice cream. On weekdays, when he wasn't wearing it, Yankel took multiple trips to the closet to admire his prize.

"Leave me over one and I'll try it," said Yankel to the diamond man in Lampertheim. If he was able to sell it, he told him, "I'll buy the others from you, too." Either the German trusted Yankel, or he was desperate: he left the stone.

Yankel wrapped his diamond in paper and put it in his jacket pocket. He borrowed a car and drove to a cutter in the German city of Idar-Oberstein. When it had been polished, Yankel could see that this stone was valuable.

Back in Lampertheim, he came upon a Greek with a BMW sports car that he'd gotten from the American army.

"Says he to me, 'Come, give me the diamond, then I'll give you the sports car.'"

Yankel's transaction was no secret around town. A man with a shoe factory heard that he had gotten his hands on a car, and approached him for a trade.

Yankel said, "I want a thousand pairs of shoes," and the man said, "Good."

"He brought me a truck—a thousand pairs of shoes—and I gave him away the car. And the shoes, I sold right away, and I earned." In one day, Yankel got rid of the entire batch of shoes. He ended up with a sack of money.

By the time the German diamond dealer came back for payment, Yankel had done his research and knew these stones were lucrative. He wanted to buy up the other five pieces of rough that he'd been shown. But it was too late; the German had sold the rest. Yankel paid the man for his one stone with food he'd gotten from the UNRRA: condensed milk, marmalade, and honey.

But there were other stones. Soon, Yankel was traveling to other cities to deal in diamonds, sometimes industrial diamonds, and eventually jewelry.

In the three years he and Dora lived in Lampertheim, they acquired a healthy capital and had their first child, a son, Steve. Then, in 1949, they received visas for the United States, where they had relatives. Yankel packed his jewelry in his bags, and the family boarded a military ship. The year before, De Beers had proclaimed "A Diamond Is Forever," but for Yankel, diamonds had nothing to do with love or glamour. Diamonds were survival.

For about a year, they lived in Kansas City, Missouri. Then they moved to New York, where Yankel went to work on the street of diamonds in Manhattan.

At the time, the district featured the Diamond Dealers Club, several inelegant exchanges, and a large café where immigrants tended to gather. Yankel didn't have enough money to join the Club, so he dealt on the street level, peddling his merchandise in the ground-floor exchanges. If he met someone on the street who was ready to make a deal, the two would find an acquaintance with a booth and carry out the transaction there so they wouldn't have to pass around diamonds in the open. "It wasn't so strict," Yankel told me. Demand for diamonds was high. During the day, he bought and sold stones, and in the evening, he went around paying for his purchases.

When I asked him what his bargaining technique was, he repeated the word "technique" and laughed, as though I'd made it up. "One says it's good, and I say it's not so good. Or one says he wants so much, say I, 'No, it's not worth that much.'"

When he is trying to buy something, he thinks of a price in his mind and then announces it out loud. He pushes that amount of money into the seller's hand and walks away, saying, "Okay, good."

He once took me shopping for a winter coat at Bloomingdale's, and to my great embarrassment, when it came time to pay, he tried to bargain there, too. "I'll give you two hundred dollars for this."

Two years into Yankel's life in New York, his brother-in-law offered him a deal: he would buy rough diamonds, Yankel would take them to Germany to have them cut, and then bring them back to the States, where his brother-in-law would sell them at high New York prices. They'd split the profits. Soon, Yankel was spending one to two months in Germany at a time. He was building an inventory of merchandise, buying and selling. During these years, he joined the Diamond Dealers Club and had a second son, my father, Paul.

The American life didn't last long for Dora and Yankel and their family. Yankel was traveling too much; Dora was alone with the boys. In a few years, they would all move back to Germany, where Yankel had made his first real money. Though they intended this as a temporary situation, they ended up staying for forty years. But instead of buying a house, they rented, as though to remain always ready to leave the country that, in their hearts they knew, had never wanted them to begin with. Sometimes, when Americans asked my grandfather where he lived, I heard him say Belgium.

Yankel remained a Club member even after he moved his family to Germany in 1960, and was still a member when he and Dora came back to New York, years later. Sometime in the early 2000s, the Club began to waive his fees, as they do for everyone who is over eighty-five.

Yankel still goes to the diamond district every Monday through Thursday around lunchtime, spending the majority of his days at the Club. Technically he is retired, but he can't stay away from Forty-seventh Street or the stones. "When one gets used to dealing in diamonds, it's like a drunkard," my grandfather says. "He drinks. He wants to drink, after all. He can't wean himself."

The Club has changed since Yankel joined in 1955. Now there are security cameras, metal detectors, fingerprint machines, and ID scanners. Now a man by the visitors' entrance waves his wand over all non-members and runs their bags through an X-ray machine. Upstairs, on the tenth floor of 11 West Forty-seventh Street, Yankel ambles into the front room, where dealers and brokers hunch over parcels at long white tables. Unlike the exchanges, which display jewels extravagantly in showcase after showcase, here diamonds are slipped in and out of pockets and passed back and forth between hands in folded papers, as though the men are practicing to become magicians. Yankel looks for people he has done business with in the past. I hear him ask them what they're carrying, what they've seen and for how much. Sometimes they take a seat across from him at a trading table and flip him a parcel paper.

The men ask Opa who I am, then they ask me all sorts of questions and tell me about their own grandchildren. But when the diamonds come out, silence takes over, and I watch my grandfather carefully unfold the package, poker-faced as he considers its white shimmering contents.

We make our way to the back of the Club, past a glass-windowed room where a Hasid weighs members' diamonds, past men with hats and men without, past the kosher restaurant, to a row of tables where Opa Yankel joins a group of men to play infinite hands of gin rummy. The Club is filled with men. There are hardly any women. But although the DDC is one of the last old boys' clubs in New York, everyone there has a girl's best friend on his mind. What Opa likes about the Club is that it is "a place where we talk only of merchandise, only of business." And though he doesn't say it, I know it's important that they talk about it in Yiddish, the district's other universal language. In many other places, Yiddish is called upon in bits and pieces, borrowed by people to make jokes. But here on Forty-seventh Street, it is serious: the language of business, spoken in full sentences, as though it exists only to put into words the character, value, and price of precious objects.

Almost every day, my grandfather stops by my father's office. I notice a shift in mood when he enters the room. He has a powerful physical presence. To stay fit, he performs a series of torturous exercises he calls "gymnastics." When I was little, I used to watch him twist his torso from side to side, letting his arms swing. Then he would sit down and beat his legs: "massage." Then he swam. A believer in the importance of circulation, he sits sweating in saunas; suffers his tea and borscht at near-boiling temperatures; bakes in the sun for hours, as though eternally trying to cure himself of those cold years in Russia.

When I worked for my father, Opa was well into his

eighties, but he looked like a newly minted septuagenarian. He stands a full head taller than his son, and my father treats him as though it were two. When Opa is in the office, my father doesn't talk very much about beauty and designers the way he does with customers. Opa picks up jewelry pieces and puts them close to his eyes, as though they were small animals whose fate he is deciding. No sentimentality. The two men talk in carats and dollars.

"Should I buy it?" I heard my father ask Opa in Yiddish.

"Buy it."

"But it's not so nice."

"Nice," Opa dismissed. The criteria he considers are objective and definable: size, cut, quality. For him, beauty is salability. As far as diamonds are concerned, my grandfather believes in things he can see.

After Yankel has tired of card games and banter and diamond hunting for the day, he takes the bus uptown or walks the twenty blocks to the apartment that he and Dora bought to be close to my family, a family almost as large as the ones in his hometown, before the war. When I was younger, we had dinner there every Friday night, and then I would sleep over so I could walk to the synagogue near my grandparents' home instead of driving. My grandfather relished the fact that I had started keeping the Sabbath. He told me I reminded him of his mother, Leah, a pious woman after whom I had been named. Sometimes he called me *rebetzin,* a rabbi's wife.

We sat around the oval wooden dining room table that diamonds bought, in a room that diamonds bought. The oil coursed through my grandmother's food like a life force, as if to attest to the fact that, despite everything, they had survived. Above us hung a crystal chandelier that erupted in colors when it caught the light.

Opa Yankel listened proudly to my sisters and me speaking in English, the most mundane chatter a feat to behold. He hardly ever joined in the conversation and rarely interrupted, as though we were scholars comparing notes. But when there was a lull in conversation, the need sometimes came over him to say, as though no time had passed between then and now at all, "A man came up to me in the camp and said, 'Do you want to buy a diamond?'" Such a simple question, a variation of the questions he and the men at the Club ask each other every day, and yet it has changed everything.

Chapter 8

A Dealer's Life

In diamonds, the passing of a family business from one generation to the next is a glacial ordeal. My father made his first sales offer when he was nine years old. He'd gotten a semiprecious agate stone as a gift from one of Opa Yankel's customers in Idar-Oberstein, the same city where my grandfather had taken his first diamond to be cut. When Opa and his next customer were finished with business, my father said to the man, "I also have something to sell you." He presented the agate stone, which was politely declined.

From a young age, my father knew he was destined to work with gems. "I always thought it was fascinating when Opa basically disappeared on diamond-buying trips for a month or longer," he said. My grandfather traveled across the Atlantic by ocean liners, so his visits to America lasted for weeks at a time. "Then he came back and showed me his purchases, and I was fascinated by these little rocks." As a child, my father would sit in Opa Yankel's home office and watch his father sort the stones he had collected abroad, carefully peering at each stone with a loupe. Sometimes my father would pick one up and look at it, too.

During high school, he accompanied Opa to New York. They traveled to the Diamond Dealers Club during the early seventies, when prices were robust. "I was sitting across from Opa and there were basically four or five brokers sitting lined up towards us at these long tables, waiting their turn, one after the other," he told me. Each time one finished, the next one moved up. On a piece of paper, my father and grandfather wrote down what they were looking for. The stones were handed to Opa, because he was the boss, but Opa passed them on to my father so that he could learn.

After high school, my father began his formal training. Every morning he awoke at six and drove for half an hour from Frankfurt to Hanau to work at a diamond-cutting factory that belonged to a customer of Opa's, to learn how to cut and polish a stone.

The first floor, where he and the other cutters worked, was perched just above ground level. Windows always remained closed so that no diamonds accidentally flew out. Lying around the room were oil jars, shellac, pliers, and other instruments. The factory's centerpiece was a long bench with cutting stations. In the bench, through six circular holes, ran six poles attached to six polishing wheels, all in a row. There were only five cutters when my father arrived. He would be the sixth diamond cutter, the last seat on the bench.

My father knew he wouldn't be a cutter forever. He knew he would join his father's dealing business. But for those few months, he lived among the cutters, working their eight-hour days, listening to the occasional jokes they told while polishing stones, taking his meals with them, and smoking together, a habit that leaves him with minty breath to this day from the Nicorette gum he chews a full decade after his last cigarette. At breakfast break each morning the cutters drank beer, always sharing with him. It turned my father's stomach to drink so early in the morning, but he made sure to take some. The cutters allowed him to feel like a coworker, though all of them knew he was just the son of their boss's friend and would soon be gone.

At the factory, my father learned to turn a natural diamond into a beauty. The course of a diamond from rough to finish is a process of refinement. The stone arrives lumpy, rugged, oily, and often opaque. It shines only slightly, the way glass shines. Not much light makes its way into the stone, and little brilliance is reflected back out.

The rough's original shape determines how it will be divided. If the diamond is a simple octahedron—like two pyramids attached at the bases—it is sawed in half by a rotating disk encrusted with diamonds, the best cutting tool. If the stone is irregular, the cleaver chisels a little indentation into the rough with another diamonded utensil, rests a knife in this groove, and taps the knife with a hammer, splitting the diamond in two.

The first steps of cutting are violent. The incisions exacted upon the rough diamond seem like the work of a butcher. But the later stages are close to the work of an impressionist painter, subtle additions to one tiny surface at a time. Except for cleaving and sawing, which are highly specialized professions, my father tried his hand in each role of cutting at the factory. First he played the girdler, inserting the unfinished diamond into a rotator and pushing another diamond against its dividing line, its girdle. When the diamonds touched, they scratched at each other, whittling one another down ever so slightly. My father held the girdling diamond on the stone he was cutting until the stone's middle was rounded out. At this point, it began to look like a diamond without defined edges.

Next, he took a turn as blocker, the cutter who applies the diamond's first seventeen facets—the facades that cause it to glimmer—and who begins the process of eliminating a stone's flaws. He put the diamond into a small shell called a "dop" and lowered the contraption onto a spinning wheel covered in grease and diamond powder, which helped shave down the stone. The blocker cuts eight facets on the bottom and eight on top, plus the diamond's actual top facet, called a "table." After blocking, the diamond looks like a gemstone with angles and surfaces, but it doesn't glitter very much yet—that is the brillianteer's job. When my father trained in this final step, he held the diamond onto a spinning wheel and turned it very slightly every so often to create the stone's remaining facets, about forty.

It took my father half a year to get to this point: to polish a diamond on his own from start to finish. He spent the first few months limited to a couple of facets on the top of the diamond, a couple on the bottom. Mostly larger, easier ones.

"At the beginning I was sort of petrified of doing something wrong." When he finally put the last facets onto his first complete stone, he beamed with pride. "Now, that doesn't mean that I cut the perfect diamond, but I cut one that had all the facets in the right order." When he finished brillianteering, the diamond was complete—shiny and shaped and ready to send out into the world.

These last facets, the brillianteer's facets, are the smallest—most of them smaller than a grain of rice—and they require steady hands to shape them. I don't think this part of diamond cutting ever abandoned my father's consciousness, because he approaches every physical task, from writing out a list to packing a suitcase, with the same steady hands, the same attention to minute detail, as though his whole world were made of diamonds.

But more important, my father's training at the factory gave him an eye for diamond cut. When he considers a stone as a salesman, he understands exactly what to look for. "I knew what it took, and I knew what a good or perfect diamond was supposed to look like, or if something was sort of out of whack." He can say with authority when a diamond has reached its Platonic ideal. Conversely, I've heard him get on the phone and complain about cushions that look like "the ugliest sons of a bitch."

Every move a cutter makes affects the way light enters and leaves a diamond, and consequently the stone's ability to shine. If he miscuts a diamond, and the distance between its top and bottom is even a little bit too deep, light begins to pour out of the stone instead of bouncing out of its top. It is lost forever, and the light show falls flat. If a stone is too shallow, it can appear dark on top, also from loss of light.

At the cutting factory, though, my father learned that the greatest danger in working with diamonds is not miscutting a stone, it's losing one. Many diamonds are as small as seeds and can easily vanish into pant cuffs or machine lenses. An arbitrator of the Diamond Dealers Club once found a missing nine-pointer (nine hundredths of a carat) in his eye. A GIA student was once handling a diamond with tweezers when the stone flew out of her grip and down her shirt. Dropping a diamond is such a familiar scare, there's a name for the position shopkeepers or dealers assume when they kneel down to investigate the floors. They call it the "jeweler's prayer."

As my father was inserting a stone into a wheel, the diamond popped out and disappeared. Opa Yankel was paying for every piece of rough my father practiced on. My father felt guilty. He got down on his hands and knees to search the floors. The other cutters searched, too. They looked everywhere in the room. It was nowhere to be found.

Finally, about an hour later, a man all the way on the other side of the bench discovered an extra stone in his tray. (Each cutter knew exactly how many diamonds were in his possession at any given time.) He returned it and my father breathed a sigh of relief.

My father's training took about two years. In addition to cutting, he studied diamond grading at the Gemological Society in Idar-Oberstein and worked at a gem-grading laboratory for six months. His plan was to take an apprenticeship at one of the largest gem companies in Idar-Oberstein in order to learn not only about diamonds but also about colored stones. But at the company, all they had him do was put price stickers on cheap animal-shaped carvings made of semiprecious gems. So he took his education into his own hands. He visited other departments in the four-story building and interviewed employees about the trade, which got him in trouble with the boss of the company.

"Well, I'm not here for cheap labor," said my father in a moment of courage. "I'm here to learn." The apprenticeship ended after about a week, and my father went back to Frankfurt. At age twenty, he was working with his father as a diamond dealer full-time.

I caught a glimpse into what their German business was like when, each summer, my mother, my sisters, and I accompanied my father to Frankfurt, where he maintained a trusty clientele. In New York, I never saw my father bring work home. I suppose this was both for our safety and in deference to the American principle of work/life separation. But in Germany everything was interwoven, and it all came together at the home office in Oma Dora and Opa Yankel's apartment, which my family referred to as the "bureau." The small room was separated from the rest of the world by an opaque glass door. While it gave off a semblance of transparency, it ultimately made you aware of that which you couldn't see.

Business was more private in Germany. Because there wasn't a society of dealers and brokers surrounding my father constantly, it was as though he was the only diamond dealer in the world. Friends or ladies of the community would come to see his collection. To get in, they had to ring up from outside of the building, and then ring again once they reached the apartment. My grandparents' front door was a heavy steel one. For decades, Opa Yankel had relied on a standard wooden door with steel panels and locks, until a boy carting a baby carriage broke in during the day and tried, unsuccessfully, to lift the safe from the office.

The ladies would say hello to Oma and Opa, and my father would come out and kiss them on both cheeks, carefully and precisely as though he had measured out exactly what point was the cheek's kissable center. Then they retreated back into the office, behind the glass door.

This was a less abstract kind of jewelry-selling than the one he usually did on Forty-seventh Street. These were real women with real bodies, and sometimes my father would have to adjust his pieces to fit them with the help of his jeweler. Like a tailor, he would study whether too much air passed between the underside of a lady's wrist and a bracelet's arc when she lifted her arm. He could have a couple of loops removed from a gold chain to bring a necklace up from the valley of her décolletage to the top of her sternum, or, if a ring slid over her knuckles with too great an ease, he would resize it or, for complicated materials, install a second, invisible ring on the inside of the main affair.

When my father was out, or when there were no customers, I was sometimes allowed into the office, a place that felt magical to me. The room was furnished with a wooden desk, a stationary bicycle, a leather couch, a marble table, and a landscape painting that included a very large moose. There was, of course, a safe and a diamond scale encapsulated in a glass case. Both of these stood in a wooden cabinet that had a lock. Other closets held enticing foreign stationery, and whenever possible, I greedily snapped up anything my grandfather let me get my hands on, as though collecting artifacts from my father's secret life. Aromatic envelopes, airmail stamps—*luftpost*—and jewelry price tickets.

My father's other German clients were located in cities such as Munich, Berlin, and Düsseldorf, so he would often wake up early in the morning to get a head start on the trip. He filled his chest pack with jewels, then drove or took the train to other places to try to unload them. Sometimes he ventured past Germany, to London and Paris. He usually came back with stickers or dolls or other souvenirs, and through his trips, he conjured up the world for me.

While my father traveled, Opa Yankel tended to the office. When he returned, my father reported on his dealings to my grandfather, who would ask questions in Yiddish. How much did you sell the diamond ring for? He wasn't interested in the sapphire earrings? Did you buy anything? The ancient language mixed with the language of gemstones sounded like the names of Jewish children: Beryl, Diamantl, Rubindl. Gems, in general, they called *brillianten.*

Back when my father was just starting out, Opa Yankel would travel with him. They worked together without hierarchy or division of labor. But Opa's customers didn't see it that way. To them, Yankel was still the head of the operation and Paul the freshman. After three years, my father set out to do something that was all his own.

It was my mother, then his girlfriend, who first got him to move from diamonds to jewelry. She was doing her Ph.D. in art history at the time and introduced him to museums, collectors, and a whole society of art friends. My father fell in love with the beautiful objects she showed him. He would drive to Paris and spend days at the flea market, buying treasures. He didn't know yet exactly how much antique jewelry was worth. As opposed to loose stones, whose value depended entirely on the balance between weight, clarity, color, and cut, artwork was more elusive, harder to price. The first piece he bought was a ring made from platinum, diamond, and emerald. He paid five hundred dollars for it and sold it for a thousand.

As a newcomer to the jewelry field, my father had to find clients. So when he was about twenty-three, he drove around Germany, hunting for buyers. "I picked a city. I drove there. I drove to the shopping district and just wandered around until I saw jewelry or antique stores, and just walked in and said, 'I'm Paul Oltuski. I'm a wholesaler in antique jewelry, can I show you anything?'"

Sometimes he had to persuade the storeowners just to let him take out his goods. "No, we have our dealers already," many would say.

"They were very narrow-minded in some places in Germany, and it took me a number of runs to be able to get in there and show them, convince them, 'Look at what I have, because you might find something that the other person does not have, and just listen to my pricing. I might be more reasonable than where you're getting your jewelry from.'"

Often he was nervous. The clients were older and more knowledgeable than he was. Sometimes they corrected him on the dates of antique jewelry pieces. He listened.

When he traveled, my father carried a gas pistol tucked into the back of his belt, always with a jacket on top. The pistol only fired gas pellets, but it looked real. In Stuttgart, while presenting merchandise to a customer at the tail end of a jewelry show, he saw a man who seemed out of place. My father had begun to examine those around him, noticing age, shabby clothing, and odd habits like lingering near other people's transactions for too long. He had noticed the man watching him, and soon a second man appeared. They were on either side of him, so my father knew he was being cased. For a short time, the pair went away. When they resurfaced, he quietly flashed his gun. The men retreated.

In a shop in Muenster, through the mirror behind the counter, my father spotted two young men loitering outside as he showed the shopkeeper his jewelry. He was already considering putting the merchandise away when the two entered. With his forearm, he swept the jewelry off the display case, onto the floor. The storeowner, an older lady, came out from behind the counter and kicked the men, shoving them out the door.

Sometimes my father's adventure stories sound a bit funnier than they are meant to. Often they feature him possessing a ninjalike awareness. But the more I spoke to dealers, the more I learned about the dangers that constantly surround them, their high odds of encountering risk at some point in their careers. I began to understand why prudence, even suspicion, was as much a part of my father's training as bartering and polished facets.

Every week, my father meets his diamond buddies at the subway stop in the district. They ride across town to a boxing gym, where they hone their skills in the martial arts. I went with him once and watched him change into a tight black T-shirt with a foreign symbol printed on its front. While he waited for warm-ups and sparring to begin, my father took a spot in front of the room's foggy mirror and stood to face his reflection. With hardcore music playing in the background, he began to move his legs, then his torso, swaying from side to side as though a spirit more agile than he had taken over. He threw punches at his image and ducked from its swings, all the while bouncing lightly on his feet. I'd never seen him like this. He moved like a real fighter, and I understood that, in his mind, he was one. Sometimes, I like to imagine that this sport was made not for boxers but for diamond dealers, who are always on the lookout for new ways to feel safe.

This vigilance: it was the way diamonds most came to define my father. Except with close friends, he avoided talking about his business the way mobsters dodge talking about theirs. He was always a private person, and I still don't know whether his discretion came from the diamonds or led him to them, but in the end, it didn't matter.

While he built his business in Germany, my father had his eye across the ocean. He longed to live in the city of diamonds and Jews. So in the winter of 1984, when my mother was seven months pregnant with me, they moved to New York City. "I remember they had all these jewels in the street and in the windows," my mother told me. "It was funny almost, because it's all so thrown together, those little *butkes,* all these totally valuable things just thrown in the window there and no nice displays or so, but just mounds and mounds of diamonds."

The district looked worn and tired. Scratches and dust clouded the exchanges' glass showcases. The floors of buildings looked as if they had been laid centuries ago. Everything was old. Even the dealers were old, for many belonged to the wave of European diamond men who had come to America around the time of the Second World War. The Club was still located at 36 West Forty-seventh Street on a small, crowded trading floor, starkly illuminated with fluorescent lighting. In the mornings, my father watched the old men eat their old men's breakfasts at the café, as the scent of herring and onions filled the room. It was a world full of men. But they were all diamond men, men familiar with the liabilities of dealing in gems, men who understood each other without having to say a word. My father felt instantly at home.

During the late eighties prices were strong, and my father was able to buy a stone from one dealer in the Club, run up the aisle, and sell it to another dealer for a higher price minutes later. In addition to newly cut diamonds, the district was home to a secondhand diamond and jewelry trade that centered mostly around the booths of the exchanges. My father was a part of this trade. During his first few years in New York, he had an associate: his brother, Steve. They worked well together and, as though to reinforce the partnership, shared a strong physical resemblance, with the same Beatles-inspired haircut and horn-rimmed glasses. Steve had left home when my father was still a boy, and started out as a social worker for the Canadian government. His funds, however, were dwindling, and my father and grandfather persuaded him to join the business. After graduating from the GIA, Steve moved to New York a few months before my parents.

In 1985, Steve and my father incorporated Oltuski Brothers, an antique jewelry and wristwatch operation. Their first office was a small rented room inside the cutting factory of a Hasidic diamond manufacturer. Each morning, Oltuski Brothers' white telephones would be covered in a film of black diamond dust. For a long time, my father and Steve believed that cutters were sneaking into their room to use their phones when they weren't in. Finally, they realized the diamond dust was entering their office through the ventilation system.

My father says that Forty-seventh Street felt right to him from the moment he first moved to New York, but Steve was not as dazzled. He preferred Toronto, where one didn't have to fight for air space, to New York's commotion and Forty-seventh Streeters' hard-hitting style of bargaining. He often spoke of going back.

There was one aspect of the business, though, that consistently delighted him: wristwatches. Steve would open them up and study their interiors, their tooth-edged wheels, hook-like click springs, and coiling balance springs. He learned to recognize each part of their entrails and came to understand how they worked. He became knowledgeable in the differences between a Rolex and a Patek Philippe and developed an eye for quality. Diamonds were hard and motionless; watches were *alive.*

My uncle was a big part of our small family during our first years in America. He accompanied my parents out in the evenings, when New York was still an unfamiliar landscape to them, and the three of them ventured out into its cafés, cigarettes in hand. But Steve was truly unhappy here. Finally, around 1988, he took Oltuski Brothers' watches (my father kept the antique jewelry) and returned to his beloved Toronto. He called his new business Northern Time. In the years to come, his home filled with watches, both whole and unfinished. He would house them in a cabinet as wide as one wall and as tall as the ceiling. In its flat drawers and in small boxes lay springs, wheels, detached faces, lone dials, and leather bands. Steve was a matchmaker. Sometimes he would acquire a watch with a bent dial but a healthy movement (the watch's mobile interior). He would keep that watch until he found another one of the same model with the opposite problem: a fine dial but a broken movement. Then he would either send the two parts to one of the many watchmakers he met over the years, or try to tackle the repair on his own.

At first, my father and Opa Yankel were skeptical of his new operation. Unlike diamonds and jewelry, watches were still an infant market. Its enthusiasts and fanatics were few and far between. But by the mid-nineties, Steve's friends tell me, he was the leading authority on Canadian Rolexes, and it was to him tradesmen came with questions about obscure watches.

Watches consumed him. When my father was in Toronto, he would come home from dinner with the other dealers to find Steve immersed in the anatomy of a timepiece, a watchmaker's loupe squeezed into his eye or a magnifying set around his head.

Steve went to all lengths to hunt down his merchandise. Once, he even partnered up with an archivist friend to track down decades' worth of employees at Toronto's T. Eaton department store, because T. Eaton awarded its workers a special model of Canadian Rolex after twenty-five years of service. Steve contacted dozens of former employees to buy their watches.

Then, in the nineties, Steve put Northern Time online. In cyberspace, before the age when precious objects were regularly bought and sold on the Internet, he gained access to inventories from the four corners of the world. My uncle's computer savvy gave him an air of sophistication. He would visit us from Canada bearing outlandish gifts, such as an electronic organizer and light-up sunglasses that looked like they came from a *Star Trek* episode. He himself seemed to have come to us from the future.

My father kept the business name, even though Oltuski Brothers had lost a brother. He found himself traveling often—across America, to Germany, to England, and, lately, to Brazil.

In 1985, his cousin had opened a purchasing office in Rio. The man knew little about gemstones, but in those days, Brazil was like South Africa of the nineteenth century, just bursting with goods. Cheap goods. You could do almost no harm in buying, but you could do better than no harm if you knew what you were looking for. My father had an eye for old diamonds, and Brazil's were incredibly economical, so he decided to join hands with his cousin. He would take the merchandise on consignment and sell it on Forty-seventh Street. He flew to Brazil every three months and stayed at his cousin's apartment for two weeks at a time. One had to stay for at least two weeks in order to get anything done. Brazilians didn't believe in promptness, and sometimes my father waited three hours or more after the agreed-upon time for a client to make his way into the office. Other times he waited for people who never materialized. He waited while people decided they wanted to go to the local beach instead of doing business. He waited while clients took ages to settle on a deal. Away from his wife and child and Forty-seventh Street, he waited.

Each morning, my father and his cousin were chauffeured to their office by a driver. A small team of hired off-duty policemen accompanied them on their way back, since it was to be expected that men would wait outside the office—what everyone knew to be a diamond office—to try to stage a holdup, or else compel the dealers to take them upstairs and open the safe. All the office employees carried guns with the exception of my father, his cousin, and the fifteen-year-old boy who did odd jobs and served coffee. Coffee ran through Rio like a tributary. My father was forced to drink cup after cup of potent *cafezinhos* while he was there—it was offensive to turn down an offer—so by the end of the day he trembled from caffeine.

In Brazil, he took chances he would never take at home. Once he followed an attorney to Bahia in a shabby airplane to see an estate owner who refused to show his goods anywhere but in his villa. The man invited my father into a dark living room to view the jewels. The shutters had been closed, and he couldn't see properly. He asked the owner to open the windows, but the man said it was too dangerous. My father insisted that he needed to see the jewelry in order to make an offer. Finally, the man relented. In the light, my father saw what the collection was: nothing. The diamonds were full of visible flaws, some gemstones were fakes. My father declined. The man, who wanted five hundred thousand dollars for the lot, pressed him for an offer, so my father said seventy thousand, knowing it would be refused. On the way back, the attorney who had taken my father to Bahia claimed he didn't know, but my father believed the lawyer was involved in the scam.

Usually, though, he emerged from Brazil with solid antique jewels and gemstones. He selected the merchandise, his cousin shipped it back to America, and both men made money. Their system continued for two years. Then the Brazilians started exporting more of their own merchandise to the United States, which increased the price of their goods back in South America. My father was no longer able to buy as cheaply, and his imported products became less unique. And traveling to Rio was no longer safe; crime had worsened. Muggings were rampant and dealers were being murdered—something that didn't happen on Forty-seventh Street. From then on, he stuck to Europe and America.

While I was growing up, I didn't know about the shady characters my father encountered, or the guns that graced his business. To me, he didn't look like an international adventurer or someone who often found himself in dangerous situations. He carried an ordinary black leather shoulder bag on his trips, which didn't appear to contain anything special.

Before he left on business, he tucked his diamond tools into the plastic sleeve of a blue travel toiletries case he got for free from Pan American Airlines on some long-ago trip. Where other men put their toothpaste, my father slipped his diamond tweezers and sample diamonds, a lineup of several stones that got increasingly more yellow from right to left. Over the years, he has collected a sample kit of nine diamonds—the colors D, E, F, G, H, J, K, M, O. He skimped on I, L, and N. When he is considering a purchase, he holds his sample diamonds up to the stone in question to determine its color.

My father no longer travels with merchandise, but back then he took necklaces, rings, bracelets, and earrings with him. All this and his personal items he put into the tattered leather shoulder bag, so as not to appear too rich, not to appear like someone who carried necklaces and rings and bracelets around. Then he packed up his family of stones and left us, his human family, behind.

Chapter 9

The Diamond Women

The first time I watched my father meet with a buyer, I was about fourteen and had been staying with a friend in Munich for a few days over the summer. My father had traveled from Frankfurt to meet me, and squeezed in some business while he was there. Beneath his formal jacket and dress shirt, he wore his chest pack, slightly swollen with diamonds. I wore no diamonds.

The office we visited looked like a renovated medieval cottage. It had a garden and was romantically dark inside. My father introduced me to the client, a German man, retrieved his necklaces and bracelets from plastic baggies, and laid them out on the buyer's desk. I tried not to move in my seat too much. I worried that if I did, I might cause a distraction that could tip the man's decision the wrong way. I understood little about the world of buying and selling, but I thought of it as a tightrope walk; any minute stir could disturb its rhythm and cause the deal to tumble to its death. This is a feeling I've never really shaken. Almost every time I accompany my father on business, I still take pains to make sure that everything is as it would be if I were not there.

But the buyer started talking to me. He asked me if I liked school. I tried to answer in short sentences so that we could get back to business. My father started to brag. He said that I went to a Jewish school. He told the man that I was in classes until five every day and that I had a double curriculum: Hebrew and English.

When his telephone rang, the man answered it and talked for long, leisurely minutes, as though we were not there. While he cupped the phone between shoulder and ear, he casually handled one of my father's pieces—a bracelet or a necklace—with his left hand, moving his thumb up and down its front side, as though petting a lizard. My father waited patiently with the plastic baggies in his lap, opaque from use.

At fourteen, he had already been an apprentice to his own father and had been on countless selling trips. But for me this was mainly an excuse for us to spend some time together. I had come as a spectator, an outsider, glancing only the peripheries of his world. Usually, he left on his own. I didn't ask about what he did, and he didn't tell me. He assumed the details of his life on the road would be uninteresting to me, and perhaps they would have been. It was only later that I wished I'd paid more attention.

The client stayed on the telephone. I wanted him to focus on the pieces my father had worked so hard to prepare and sell in his own quiet, not pushy way. But my father sat good-naturedly, as though he would be graded on behavior, all the while balancing the plastic baggies in his lap. At one point, he asked me what I wanted to have for lunch, as though we were just on a sightseeing trip. I told him I didn't care. It felt

as if lunch itself depended upon this buyer. He was the first customer I'd ever met who wasn't also a family friend.

In the end, the man decided he wasn't going to buy anything. With all the care I'd taken for decorum, I'm not sure it had occurred to me that my father's jewelry could be rejected, and this revelation pained me.

My father put the necklaces and bracelets away quickly. He slid his fingers under the middle of each piece and lifted it so that it hung flaccid from both sides, like a little Dalí clock. Although he spent hours in advance of a trip or show packing his jewelry and diamonds in ways that prioritized order, safety, and the well-being of his merchandise, my father could apply an impressive agility to cleaning up. He put away his things and thanked the client warmly on our way out.

We got our lunch in Munich's city center and sat eating it under an expansive café umbrella that covered us in a grid of shade, while the sun shone over the rest of the city. It had been very strange to witness my father as a salesman.

Usually, I was on the other side of the world, or at least of Germany, when he went on these trips. While he was away, I always kept track of what time it was in Europe. It was as though one of the miniature alternate time zone clocks embedded in the circle of his wristwatch had transplanted itself into my mind. Sometimes he called me before I left for school, and my mother brought the cordless phone from our kitchen into my bedroom at seven thirty a.m.—my wakeup call. Still dazed with sleep, I would listen to him very carefully, prizing our time on the phone.

If I had thought I would be heading down the same career path in a few years, I probably wouldn't have worried so much when he didn't sell a piece to a client, or when he was away on the road. I would have understood these things as a natural part of the business. But business and money were matters my parents kept at a discreet distance from their children, except for the few times they let slip that things were not going well. Sometimes my mother, wearing two or three diamond rings on her hands, would bemoan how expensive kosher meat had gotten. In the next room our cat, Diamond, snacked on Fancy Feast.

The jewels themselves made frequent appearances throughout my childhood, but the particulars of my father's diamonds were as hidden from me as they were from strangers. I didn't even know what to tell my friends when they asked what my father did. My father's answer when I posed this question to him: he was a businessman. I knew enough not to equate diamonds with wealth, but it was hard to decipher at any given time whether we were rich or poor or somewhere in between. I can't help but wonder if all this would have been different had I been born a boy. Would I have been let in on the state of our family trade? Would he have told me how much he'd paid for a necklace and how much he'd sold it for? Which clients to avoid and which to seek out? My father claims no, and in all fairness, I never asked him for details. But I think I might have, if I'd imagined myself the successor to his affairs.

It's obvious that I'm not the only girl who, for one reason or another, didn't end up following in her father's diamonded footsteps. For a long time, the diamond business was just about as masculine as the Wild West, and this is changing only slowly.

In commercials, diamonds make beautiful men and women fall in love a thousand times a second, their chemistry as aflame as the white rock resting in the boyfriend's palm, on the wife's neck, afloat on the ears of a female shadow. The closest I've come to seeing that kind of romance on Forty-seventh Street is the occasional couple shopping for an engagement ring. I stopped a pair of lovers to ask why they decided to come to the district, and got the predictable answer: they knew someone here. Meaning, they would get a good price. The couples I see most often are pairs of men. They are talking business, making deals, sometimes just schmoozing.

The street is not a very sexy place. I don't feel particularly womanly when I'm there. The catcalls tend to disappear as you enter the district. You're paid attention to—you, as a female—but it's not your breasts, hips, or backside they're looking at. It's a select few of your fingers, maybe your earlobes, and your neckline. Most of all, you are interesting for the material desires you're supposed to possess, for the fact that a diamond's shimmer is supposed to run electric currents so deeply through your body, you mistake it for love or something better.

I know the look by now. I've been to the exchange many times, and if it's not a busy day, the eyes of booth holders follow me and my gaze around the room, wanting to know where I'll stop, what will strike my fancy. They have already analyzed me to predict whether I am in the market for a pair of earrings, maybe a necklace. I am not there with a man, so I'm probably not buying an engagement ring. I don't look very rich, I don't even look casual in the way rich people look casual, and I'm young, so I'm not likely to spend more than a couple thousand. I may be a working woman with a bonus check to fritter. I may just be browsing, wasting their time. But there is potential in these women who wander in.

At the Club, I am in a world filled almost entirely with

men. It's not that women are unwelcome. It's just that there aren't many there, at least not on a regular basis. But some come. One of them has been around since the late seventies, when women's membership was still a novelty. Every morning, at around nine thirty, she takes her seat at the first trading table on the left side of the room, by the window. There Edith Lipiner, Sal Lipiner's wife, has the double benefit of direct daylight and a full view of anyone entering the Club. She sets up her scale, lays out her parcel papers, and waits for business to begin.

I met Edith up front, by the guard's quarters near the turnstiles, and I felt something I hadn't felt with most of the male diamond dealers I'd been spending time with over the past few months: intimidation. Her gray and blond ringlets were pulled back tightly. She wore a skirt suit. She looked corporate and had the confident air of someone who knows her time is valuable. While we spoke, we were interrupted several times by greetings from other members, the Club's paging system, and Edith's own cell phone. "Hello? Yes. Yes. Thank you very much, and how much is it?" I don't know why I thought there was a chance she would fill me in on the deal she was talking about, but when she didn't, I knew I couldn't ask.

The jewelry she wore that day was subtle: a gold bracelet, a diamond ring, and on her wedding finger, a white-gold band sprinkled with upright diamonds.

"Do you ever wear your own pieces for advertisement?"

"No. I don't believe in that. Because I'm not in jewelry. I'm in diamonds, so it's different, too. But I know a lot of women will even wear earrings or different things and then sell it right off them—I don't do that. I do wear the diamond studs, because I think it makes women say, '*Mmm,* diamond studs. That would be a nice thing to have.'" I didn't mention that my father sometimes dressed my mother in his fine merchandise when they went out to fancy events, in the hopes that some rich acquaintance would be tempted to buy.

Edith's seat near the front of the Club used to be her father's. Back when the new hall was being built, a committee member took them up to see it and pointed to the first table on the left. "And that's going to be your table, Myron, because you could see people as they walk in." Edith's father died a few weeks after the new Club opened in 1985, and Edith has kept the place ever since. Now it is she who looks at the people as they come in.

She wasn't always an established Club member, though. Getting settled in the business was not easy. The resistance first came from within her own family. She was in her thirties when she approached her father and uncles and asked to join their business. She felt that someone in her generation ought to know their way around the family trade, "but they were appalled, because I was a girl, and they just couldn't understand."

Then, in 1979, one of her uncles died, and the other had a heart attack during the mourning period. Edith went from unwanted to indispensable. She became her father's trusted assistant, accompanying him on business trips to Belgium multiple times a year. But Myron was still ambivalent about her working with him, because she was his daughter, not his son. When Edith balanced the books, he wondered, "Maybe you should be looking at diamonds and learning about diamonds." When Edith handled the gems, Myron was prone to nervously contradicting himself. "Well, you really should be making sure all the books are settled, all the paperwork is done."

Sometimes he would say condescending things. One day they both got into the elevator in 36 West, back when the building still housed the Diamond Dealers Club. In the elevator were two other women, one a blond beauty. Edith and the women said hello.

"Isn't that nice?" said Myron. "All the secretaries know each other."

"I had a real terrible feeling, when he said that," Edith told me. "And I thought to myself, 'Well, that's sexist.'" It was also inaccurate. The first woman happened to be married to a successful Forty-seventh Streeter. The blonde was employed by "a man who was considered"—Edith paused—"a gigolo." She and her boss were known to go on frequent business trips around the world. Edith tucked the incident into the back of her mind until several weeks later, when she came across a newspaper article about the globetrotters. As it turned out, the blonde was a Czechoslovakian spy, traveling the world to carry on her political affairs and using an oblivious diamond dealer as her free plane ticket. But all Myron Sokolik had seen were three typists in an elevator.

He was "the kind of person who came home and did not talk about business," Edith told me in the Club that day. I recognized this. My father never spoke about his days on Forty-seventh Street either, or about the deals he made or the men and women he got to know. When I was growing up he didn't tell me about the striking gemstones he held in his hand. The thirty and forty caraters, as large as eyeballs; the chameleon diamonds that were green in hot weather, yellow in cold; or the rare blue diamond he was once shown in the office of a prominent dealer. The stone was so beautiful that it is still imprinted on what he calls the eye of his mind decades later. Never again has he seen a diamond so blue.

The Club was often a lonely place for Edith. Sometimes she wished she were a man. She had little to do with those few other women who frequented the trading hall. They were much older than she was; they were the widow members. Until 1978, a year before Edith entered the business, females were not allowed actual Club membership, but some rights were granted to the widows of former members, or the wives of men who could no longer work due to injury or poor health. They were allowed to use the Club's facilities, to trade in its halls and be paged on its loudspeakers. But because they were not considered real members they could not vote in elections, nor were they charged a full-member fee. The point was to give them an opportunity to salvage their families' livelihoods.

In 1978, when the first female member joined the Club, the widows became full-fledged members, which meant full membership dues. Although a number of them were unhappy with their costly promotion, Sal Lipiner told me, they had no choice; when the Club began admitting female members, it abolished the post of non-member trader. Since most of these women were brokers rather than dealers, they didn't all come on a regular basis anyway. And though Edith held the occasional conversation with them, they were not her close friends.

Neither were the wives of her diamond colleagues with whom she sat when she and Sal were invited to the religious weddings of Forty-seventh Streeters, where there was no mixed seating. "It was interesting," said Edith. "It was very interesting. Sometimes they told me things they shouldn't have told me." One of the wives revealed whom her husband was selling his diamonds to. It was a customer of Edith's. Edith could have cut the husband out of her dealings and sold straight to his buyer. But she wouldn't have done that.

She wasn't one of the boys either. In the old days, women across America fought for a seat in the boardroom. At the Club, Edith fought for a space in the chapel. When the new synagogue was being built in the trading hall, she told the president that there should be a divider so that women could pray, too. "Well, they laughed a lot. And then he said, 'Okay, we're going to do that.' Of course, they didn't do it. And if I ever say that to anyone, they'll say to me, 'Well, would you really go in there and pray?'"

I've been to the chapel a few times. It's a room at the back of the Club filled with books. Aside from the buzz of prayer when the men congregate for *mincha,* it is a quiet room. One or two men sit at a table with an open Hebrew text, studying. On the back wall, an electronic lunar calendar broadcasts whether the rain prayer season has begun and what time the sun sets that day. No one has ever asked me to leave that room, but no woman has ever joined me in it. The solemn feel of scholarly men presides there, as though it were a monastery or a yeshiva.

A year after she first came on board as her father's associate, Edith applied for Club membership, as the daughter of a dealer. This meant that her father pledged to sponsor any transactions she made. Dealers often worry that young new members won't have the capital to pay for the diamonds they buy, that they'll promise a check in four weeks, then default. A father's guarantee gives them peace of mind. Because Edith's family was established in the trade, her interview went swiftly. No one had any doubts about her. All that was left was protocol: Edith's picture had to hang on the Club's admissions bulletin board for two weeks to give the membership a chance to contest her joining, during which time, to Myron's distress, Edith was barred from the Club.

The bulletin boards are the DDC's way of disseminating news and information, both good and bad. I once saw a man's picture pinned to one of the boards with a note explaining that he hadn't complied with a ruling of the Club's arbitration tribunal. Other boards serve as the equivalent of classified ads for diamond dealers: "semi-retired man looking for a position in diamond district"; "1-1.33 ct. G-H IF" (a desired stone); "GIA-CERTIFIED 1.23 ROUND / FVS1" (a lost stone); information about the wake, Mass, and burial of a member; and a warning note from a company stating "Criminals are asking for goods in our name."

If the complaints against an applicant on the bulletin boards are serious enough, the Club will deny him membership. But of course, no one had anything bad to say about Edith. They knew her face, and more important, they knew her family.

Not everyone is this lucky. Jisun "Sunny" Chung, one of the other regulars in the Club, had to wait thirteen years to join—the cost was prohibitive, and she didn't know enough members to guarantee her the first two times she tried. But Edith's two weeks passed without anybody protesting her entrance, and she became a full-fledged member of the Diamond Dealers Club. A few months later, her other uncle died.

Despite her lineage, Edith isn't a particularly outspoken constituent. She doesn't even take her meals in the lunchroom, "because the men don't"—she lowered her voice, for we were among the men—"the religious men don't like you to sit and eat with them." For reasons of piety, most of her colleagues won't shake her hand. But this isn't something that bothers her. The Club is a place of faith, and Orthodox Judaism has become the unofficial law of the land, even for those who are not part of it, sustaining a delicate ecosystem of Jewish men. Sometimes there is a divide between men and women. It's not one people condone, but it's there.

Susie Ehrman milks the divide. Being a woman in such a male-dominated business has brought her nothing but good fortune. I met her in her small office on Forty-seventh Street. She is actually a distant cousin of mine, related by marriage to our diamond-dealing relatives in Antwerp. Back in the old country, the story goes, the Altuskis had called themselves Oltuskis. Then a woman in the Oltuski family scandalized her people by becoming a stage actress. To separate themselves from the heretic of the family, the religious contingent changed its name to Altuski. I stem from the errant side of the family, Susie's relatives. I don't even want to think of what the old Oltuskis would have done if they had had to deal with female diamond merchants.

But it was being a woman that got Susie her most important gig. It got her a job with Martin Rapaport during the early days of the price list.

Susie grew up just a few blocks from Antwerp's famous diamond district, where men would walk around with their diamond purses chained to their suits. She came from three generations of diamond-dealing Ehrmans. Their last name meant "man of honor," and their reputation in the business matched it. They could get credit from anyone they wanted. But, like me, Susie had no soft spot for the family trade, and her father never trained her in its particulars. She was supposed to become a doctor of psychology. She described herself at the time as "antijewelry, antidiamonds," but added in a murmur, "I caught up really well." She was wearing flower-shaped diamond earrings when I met her.

After a semester of graduate school, Susie got engaged and followed her fiancé back to Europe. When she broke off the engagement, she returned to the States but found she had already missed too much of the semester to reenter the program that year. So she searched for employment.

Then one day she met Martin Rapaport. Back then he was called Mike. He was out Rollerblading in shorts with a friend of hers on the Upper West Side of Manhattan when she made his acquaintance. Susie thought he was crazy.

When Martin discovered that she came from a diamond family, "he got all excited."

"Call me tomorrow," he said. They met in his office, a small space he rented from an elderly diamond cleaver, and he asked her to work for him.

Susie remembers Martin as an agitated eccentric who spoke "a mile a minute" but dutifully molded her into an educated diamond broker. He sponsored her six-month course in gemology at the GIA. "He showed me rough diamonds. He showed me polished diamonds. He showed me how you close a cachet—*everything* there is to know."

Martin saw the scarcity of female workers in the district as an advantage. "There's no women in the diamond business, and because of your family and your family ties and your strong name, I think you'll be an asset to me." And she was.

"Send Susie, send Susie. She's more fun," begged the dealers with whom Martin dealt before he began publishing prices.

"They liked to have a fresh face."

Her fresh face would be even more valuable after Martin started the price list. "You know, Susie," he said to her one day, "you as a woman, 'cause you're attractive and all that, you'll go downstairs. We'll set up a booth." *'Cause you're attractive.* I liked that Susie didn't put on false modesty. Martin liked this about her, too. He figured that at least the diamond dealers would look at her, and he was right. While Forty-seventh Streeters heckled him incessantly in the months following the publication of his list, shouting things like "You're gonna ruin our business. You're gonna destroy us," they never bothered Susie. Of course, Susie did not go about her duties in the inflammatory way Martin did. She didn't stand in the lobby of the Diamond Dealers Club, the epicenter of price list resistance, distributing the list and calling out to passersby, as though she were selling the morning news.

Instead, her job was to deliver the list to major offices on Forty-seventh Street that were friendly to the idea of the report. Her other responsibility was to phone a few people each Thursday and ask them how much they'd sold their diamonds for that week, "and that's how he made sort of an average. Do you understand?" Susie asked me.

I couldn't believe what I was hearing—the mysterious source of Martin Rapaport's prices. So I asked again.

"I'm telling you," said Susie. "I used to call up the big offices, and I said, 'What did you sell this week and for how much?' and that's how he sort of made an average, so he got some examples of how much things were sold for that week. Do you understand?"

"So how many companies did he call up?"

"Between five and ten, maybe. Not a tremendous amount,

to tell you the truth." And there it was, the information that had left so many people guessing for all those years.

Susie was there when it all started. She was there when Martin's supreme influence on the industry was only a dream. "You'll see, my name will be on that list, and everybody all over the world will know my name," Martin announced in the office.

"He was saying it almost every day, almost like a mantra, and everybody thought he was crazy. His name all over the world. And he fulfilled that dream of his. His name is known all over the world."

Their relationship was a symbiotic one. Through Martin, Susie not only gained experience in the diamond industry, but she became well known on Forty-seventh Street. Martin encouraged her to join the Diamond Dealers Club and guaranteed her membership in 1980. In return, she helped him collect the data for his list, distributed it, and acted as a representative for his brokering company.

Susie found the overwhelming ratio of men to women in the district could serve to her advantage, even after she stopped working with Martin. "I liked it, because I stood out." When she called a Forty-seventh Street merchant with whom she hadn't previously dealt, she would say, "Maybe you know me." She would mention her long blond hair, and they would say, "Yes, I know you by sight." That wouldn't have worked had she been a man with a beard who wore a black hat and black coat to work every day.

But the same femininity that gave her an edge also had its disadvantages. Following her employment with Martin Rapaport, she was recruited to become the New York director of a Utah-based investment firm called the Gold and Diamond Exchange, which she left after about a year. Diamond prices had crashed, and Susie wanted to try to make it as a private dealer.

One day she got a knock on the door of her Forty-seventh Street office. Through the surveillance camera, she saw an outstandingly tall man. She buzzed the doorman of her office building. "If I call you in ten minutes," she said, "I'm okay. But if you don't hear from me, come and get me."

The man was from Guinea, where the Gold and Diamond Exchange had invested in a mine. He said he knew Susie from the company. She didn't recognize him, but she knew that her name had appeared on the firm's roster. I try to imagine what I would have done in the situation. I probably would have asked for a meeting in a public place. But then again, diamond people don't really do public places. Whether bravely or stupidly, Susie let him in. "And he comes in, and he opens his briefcase"—Susie adjusted her voice to one of quiet wonderment—"and he shows me parcels of diamonds that people dream of, and he said, 'I want you to help me sell this.'" The rough diamonds were as clear as glass and large as blueberries. A dealer's mirage. Susie knew she didn't have the capital to purchase them on her own. She would need partners. So she called up one of her father's friends on the street who bought and sold rough. The dealer invited some of his friends. They all met in his office—Susie, the Guinean, and the Forty-seventh Streeters. "They looked at each other when they opened the briefcase. They couldn't believe what they saw."

The diamond men inquired as to the origins of the gems. The Guinean said they came from Guinea but wouldn't give any further details. The dealers overcame their wariness enough to buy $500,000 worth of rough then and there. After the transaction, they escorted him into another room of the office and, breaking diamond etiquette, left Susie behind. Susie was too stunned to protest. They paid her the standard one percent commission for making the liaison between them and the Guinean, but she later found out the men had made subsequent deals with him behind her back, locking her out of the profit.

"I was twenty-three years old, on my own, and they just took advantage of a young woman who was putting the trust in them and that was it. Since then I learned that never again will I introduce anybody to anybody. Ever." Had she been her father's son, rather than daughter, I think there's a good chance they would have let her in on the action. But Susie survived. She's still working, now as a jeweler rather than a loose diamond dealer. She prefers selling to privates instead of dealers—it's easier to get them to pay on time.

Sitting in her office on Forty-seventh Street, I was fascinated by her drive in an industry she used to consider boring. It was happenstance that brought her from psychology to this business. Almost none of the diamond women I met were destined to go into diamonds. They were the exception—anomalies in a neighborhood that exists to dress women in jewels but that is almost devoid of them in its Club, in the boss positions of its upper offices.

How much easier it is to float around the district as a customer, to walk up to a booth and, with a word, command knots of gold, silver, sapphire, emerald, and diamond from beneath the glass, as though your lips themselves were magnets. Having heard Edith's, Sunny's, and Susie's stories, I can imagine some of the encounters I would face if I ever chose a life on the other side of the booth. I know that I would be entering what is still very much a man's world. Sometimes, though, I still wonder, if my father had asked me to come on board, what I would have said.

Chapter 10

The Last Diamond Cleaver

Tucked into the backs of Forty-seventh Street offices, deep within the buildings of the district, are where the diamonds get cut. Most of the factories are operated by Hasidic men. Unlike many other American plants, they do not inhabit vast landscapes on the New Jersey Turnpike or lonesome rural terrains. Unobtrusive, they look like any other diamond workplace, from the outside, at least.

Efraim Reiss took me past a door that separates the cutting quarters from the rest of his father's office, past the counter where the men eat their lunches and pray, into a medium-sized room with stark white lighting.

The factory was a loud place, and every once in a while a stone screeched when it hit the wheel. The area buzzed lightly. Almost everything in the room—the ceiling tiles, the walls, the hands of the cutters, the Post-it notes, the FedEx envelopes stuck in between the two rows of lighting above the cutting wheels—was tinted black or gray by diamond powder mixed with oil and grease. Diamond dust does not look very different from regular dust, except that it sparkles faintly. My second time at the factory, I asked Avi, the cutter showing me around, why the white stones produce black dust. Avi's hands were calloused. They were so worn from the wheel that his fingerprints were topographically raised above his fingers, visible to the naked eye.

Because, he answered, it's coal. On one wall of the factory hung a small wooden plaque that said *A Diamond Is a Chunk of Coal That Made Good Under Pressure.*

Below the sign was a computer. To check the evenness of their work and plan their next move, the cutters insert a diamond into a box that's hooked up to a program, and the diamond's body, highly magnified, appears on the screen. They can see a diamond's angles and symmetry, its height and width. An onlooker can also see the cutter's fingers, chafes and all, descending upon the stone.

Near the entrance to the factory was a small dark room where the resident gemologist graded diamonds. An alcove nearby housed the diamond cleaning station and girdling machines. When the girdler, an Israeli-born man named Shmuel, flipped a switch, the machine, a hefty box, lit up. Inside, two diamonds on two separate dops resembling silver bullets twirled around each other. Through a small window, I saw the shadows of the diamonds. A mirror inside the machine reflected the stones. In a diamond factory, Shmuel mused, "every drop is money." He told me that girdling produces more diamond dust than cutting.

Opposite the girdling machines was Shmuel's worktable, on top of which stood a toaster oven, used to heat the glue that binds the diamonds to their dops; a Sudoku puzzle book; a collection of Hebrew psalms; and a charity box. Above the table, a glass box with two small doors was mounted to the wall. It housed a fan, a ventilation pipe, and two burners. I watched Shmuel put a heart-shaped stone in a container of yellow liquid on the burner. At one point, I asked him how much one of the diamonds was worth. He told me that wasn't his job.

The majority of the factory's space was devoted to its fourteen wheels and their accessories. Counters ran along four walls, including a partial one that divided the area in two. On top of the counters were the cutting wheels, metallic circles a little smaller than medium New York pizza pies. Avi let me put a finger on a spinning wheel, then lifted it off a moment later to show me my finger covered in diamond dust. The cutters and the girdler recycle leftover dust they produce, spreading it onto the cutting wheels.

All around the room—on the tables, affixed to wheels, hanging from metal shafts on the walls like wineglasses at a bar—were dops. There were red dops and silver dops, dops used for cutting the tops of diamonds and dops used for cutting the bottoms of diamonds. Sometimes three, four, even five dops can be attached to a wheel simultaneously.

The first time I visited, I noticed that most of the cutters sitting at the wheels wore headphones. Torah tapes, one cutter explained. Beneath one of the tables was a box packed with audio cassettes. As the factory fills with the drone of diamond cutting, the workers' ears fill with the sounds of Jewish lectures. In Jewish tradition, the Torah itself is seen as a gem. Except for Avi, who is Sephardic (a Jew of Spanish or Portuguese origin), all of the cutters are Hasidic.

I'd brought a camera. A man sitting at one of the wheels in the back covered himself as I took a picture. Many diamond men refuse to identify themselves to the public because they worry about safety. But the second time I came, most of the factory staff opened up to me. They volunteered information, like how slow business was, how hot a dop could get, and that, in Israel, they have special wheels that are already coated in diamonds, disposing of the need for diamond-powdered oil. Some of the men offered to let me touch their wheels, and I felt their smooth interiors, which enable the finer task of polishing, encircled by their rough sandpaper-like outer rings, for cutting. One cutter, a slim, dark-eyed Hasidic man also named Shmuel, approached me to show me a rough stone marked up with ink, indicating where it would be cut. He let me hold the stone in my hand. When I'd finished handling its smooth but uneven body and wanted to give it back to him, I realized that I shouldn't simply take the stone between my fingers and place it into his palm, that this would likely result in our touching hands, forbidden for him. Instead, I left the diamond in my palm and positioned my hand slightly above his. I began rotating my hand from horizontal to vertical until the diamond dropped, and he caught it without breaking a single one of his laws.

Sometimes a stone moves back and forth between different cutters in the factory before it is finished; although the men all do similar work, several of them have special expertise, like diamond shaping and polishing. Avi can work on five rough and two polished stones at a time.

Before he completes his duties, each cutter leaves a little imperfection on the stone to show the boss, Jack Reiss. They call this the *baveis,* the Yiddish word for "proof." As *baveis,* Avi will leave a small crack in the diamond to validate that his incisions were absolutely necessary and did not extract valuable weight from the stone.

At first, it surprised me that a job that relies so completely on the skill and trustworthiness of its workers would resort to such an invasive form of safeguarding. But diamond cutting is still a subjective art form. The factory employs twelve cutters, two girdlers, and a gemologist, but it is always Jack Reiss who marks up each rough diamond with ink to direct the shape the stone will take. While I was there, Mr. Reiss poked his head into the factory at least twice.

Even though *baveis* is par for the course, the pressure on cutters has been mounting, Avi told me. With the growth in diamond-cutting computer programs, which can calculate a stone's exact dimensions, buyers insist upon complete symmetry, exact angles. In most American cutting houses, human hands do the cutting but computers judge the handiwork.

Most cutters on Forty-seventh Street are self-employed and earn their income from several bosses rather than have a steady salary from only one. They get paid per carat, which can range from about fifty to five hundred dollars per. The diamond cutters in Jack Reiss's factory are allowed to keep their own hours, since they work on a diamond-by-diamond basis. As long as the stones get finished, the cutters can come and go as they please. The majority of them commute in on the bus from Monsey at around ten or ten thirty, but Avi arrives at seven thirty or eight in the morning, and he stays until seven at night. Sometimes, he told me, he dreams about diamonds.

Before diamonds even get to factories, some of them are ministered to by Ben Green, the self-named last diamond cleaver of Forty-seventh Street. He is a big man for his modest office. Not big as in plump, but big as in strong, monumental, so you can see that his work is of a physical nature—never mind that the stones are delicate and sometimes smaller than his knuckles—and that he is cut out for the job. He welcomed me in and asked me to sit down and told me my father was a special man. I said I knew.

Mr. Green says he has been the last diamond cleaver for over two decades now, and even he has invested in a diamond laser, since stones are rarely cleaved anymore. He is one of only a few men in the district who own the machine. Though cleaving is viable only along a diamond's grain, and sawing only against the grain, lasering provides flexibility; any orientation of cut is possible. Mr. Green compared it to an airplane, which can fly in any direction, as opposed to a car, which must stay parallel to the ground. But every once in a while, he comes upon "a very, very dangerous stone" that demands manual cleaving because of a fracture in its interior. If the laser hits the fracture, the diamond can explode. Despite advances in technology, the risk involved in splitting a rough stone in two has not disappeared.

I asked Mr. Green if he ever gets nervous. In return, he asked me if I thought a heart surgeon can perform when he is nervous.

"Probably not," I said.

"Not," said Mr. Green, a round-faced, goateed man, dressed formally in a pin-striped suit and a tie. He wore snazzy half-frameless glasses one doesn't usually see on a man in his seventies.

Mr. Green left Europe for Israel illegally after the Second World War, along with three of his siblings, when he was thirteen. Four hours away from its destination, the boat he'd boarded was taken over by British forces and its passengers detained on the island of Cyprus. There, the British locked the captives in a camp that held seventy-five thousand people. At night, they slept in tents, with five to thirty-five people to a tent. Three months later, a United Nations program took him and the other children to Israel, where Mr. Green attended high school at a yeshiva.

He was in his thirties when his brother, a diamond manufacturer, introduced him to the business. After training as a cleaver, he came to the States in 1973, during the heyday of New York diamond manufacturing. Today, while large and fine diamonds are still cut in New York, the district has become more of a dealing hub than a cutting depot. But back then, it was one of the world's major polishing centers, and just about every irregular stone in the city was cleaved by one of two old men. Less than a decade after Mr. Green arrived, the men retired, and his business skyrocketed. Mr. Green revealed to me the carat weight of the largest stone he's cleaved, but then asked me not to repeat it, because it is a "secret of the business." If the gem belonged to him, Mr. Green wouldn't mind, but etiquette demands that a cleaver not speak publicly about the size of other people's diamonds. When I visited his office, though, pictures of his conquest hung on the walls. It was enormous, maybe the size of a baby's fist.

The rest of the office was crowded with tools—small glass jars with alcohol and water inside them, a knife sharpener, a Pyrex pitcher, a fire extinguisher that he's never had to use, two boxes of Huggies wipes, a funnel, a scale that can weigh diamonds up to two hundred and fifty carats and another that can weigh a thousand-carat diamond, Comet disinfectant powder, a glass teacup with a bit of water inside,

Wite-Out, and two radios—Mr. Green listens to Israeli and Hasidic music while he works, though he is neither Israeli nor Hasidic.

"And the fridge is just for food, not for diamonds," I asked.

"No, for food. For diamond, this fridge is for diamond," he said, pointing to the safe.

Mr. Green rents two offices, each from a different friend in a different building on Forty-seventh Street. In one of his spaces, he keeps his laser machine; in the other, where I met him, his traditional cleaving materials. His work begins over a small, hollowed-out block of wood built into his table. There, he creates a groove within the rough diamond, where his cleaving knife will rest. The key to cleaving is a good groove. It must be made with precision. A millimeter too deep, and the diamond can shatter. To make the little trench, he uses a second diamond, but instead of holding the two stones in his hands, he wedges them into implements called cleaving sticks, which look like fat paintbrushes wrapped with clumps of cigar paper. The sticks, I learned, are one of the few instruments of the diamond industry that have not changed for centuries. About a dozen of them lie around Mr. Green's desk. The clumps of cigar are actually a special kind of cement made for cleaving. In its natural form, it resembles shards of pottery. Before fixing a diamond into the top of the stick, Mr. Green holds the cement in an open fire emanating from one of the small glass alcohol burners that populate his table. When I was there, he took out a lighter with a Yankees logo on it and lit a burner. Then he heated the cement and scratched it with a knife to demonstrate.

When the cement is soft, Mr. Green presses the diamond in, with the half that will receive the blow faceup.

Then he cuts the groove. Small shards of diamonds fall into the wooden box, but they don't go to waste. Mr. Green sells them to cutting factories, where they'll be used to coat their wheels. Not every bit of diamond can be salvaged. Often he gets the residue on his fingers, and some of it graces his worktable, as though the office were showered in fairy dust. The price of diamond dust has gone down over the years. When we met in early 2009, it ran for under fifty cents a carat, a big comedown from the seventies, when the going price was five dollars. The decline is partly due to the availability of synthetic diamond powder.

The actual cleaving takes place beyond the table. Affixed to the brink of the desk is a little ledge with a circular hole in it. After he makes the groove, Mr. Green slides a stick with the rough diamond into this hole and pushes it firmly into place. He fits a metal knife in the groove and lightly taps the knife with a hammer. The insides of a diamond are organized in sheets of carbon crystals, so the diamond splits along internal lines. One half of the stone stays wedged in the stick's cement. The other falls onto a green cloth that's secured beneath the table, there for the catching. When he is done cleaving, Mr. Green extracts the cemented diamond the same way he inserted it, by holding the stick over an open flame.

The cleaver is the biggest gambler in the entire cutting process. If the knife strikes against the sheets rather than along them, it can shatter the stone. Some say that Joseph Asscher, the man who cleaved the famous Cullinan diamond for King Edward VII, fainted after he'd successfully completed the job. That was in 1907, and a larger gem diamond has yet to be unearthed. But stories like these don't faze Mr. Green.

From start to finish, cleaving can take up to ten hours. Timing is subject to how "dangerous" the stone is, its size, its inner structure, even what mine it came from, since its origin affects the way its atoms have bonded, its impurities. When I asked Mr. Green whether he's ever made a mistake, he answered in the negative. But even his skill won't bring back the art of cleaving. Soon, all his strange instruments will be artifacts. Lasering, which can take just as long as hand cleaving, will be the only way to split a diamond in New York City. "I am almost almost retired," Mr. Green told me.

Over the course of his career, he has seen much of his work get outsourced to India and China, where labor is cheaper. When the economy began to decline, work became even less steady. Mr. Green never knows when his next job will come. "The last stone that I cleaved, it was two weeks ago. That was a ten-carater stone. Ten-carater stone, that means a small stone."

"Really?"

"Yeah."

"Wow," I said. "For me, that's a big stone."

"For you it's a big stone. For cleaving it's a small stone." Mr. Green's going rate is usually twelve dollars per carat. The most he ever charges to cleave a diamond is twenty-five dollars a carat, and that is only for very large diamonds. A ten carater could take one or five hours to cleave but would earn him a total of a hundred and twenty dollars either way. For those kinds of stones, "here in New York, there is no, I will call, no *parnasa.* You know what's *parnasa*? There's no living." Mr. Green decided not to pass his business on to his son.

*

The craft of diamond cutting owes its existence to a series of social transformations hundreds of years ago. Diamonds were not always thought of as malleable commercial goods, there to process, improve, and sell. They were far more important than that.

For much of their history, the gems were elevated to near divine status. The Greeks supposed they came from the sky; diamonds were star slivers that plummeted to the earth. Romans, who believed the gemstone had magical qualities, wore diamond rings for protection.

Because of their supernatural associations in pagan religions, with the rise of Christianity in the Middle Ages, diamonds were essentially banished from Europe for almost one thousand years. But over time, diamonds left the realm of the sublime and settled into the role of luxury items. By the thirteenth century, they decorated European insignia and jewelry. They were used widely in the wedding rings of the rich. And by the fifteenth century, the diamond's durable physical properties had turned it into a sign of marital faithfulness.

Then something happened that would transform the stone's role forever. In 1477, Archduke Maximilian of Austria gave his betrothed, Mary of Burgundy, a diamond ring. Although promoting love had been one of the gemstone's supposed powers in older times, Mary's is the first known diamond engagement ring. Soon, European patricians were heralding their nuptials with diamonds. So diamonds went from being objects of might and magic to emblems of romance. Everyone who could afford them wanted them, and over the years, craftsmen unlocked the equations of angles in a stone that would make it glow brighter, giving birth to the vocations of men like Mr. Green and Avi and the Reisses.

We don't think of diamonds as godly anymore, and yet on Forty-seventh Street, they seem religious to me. The Torah tapes, the sound of *Mazal* every time a stone changes hands, the prayers that fill so many of the neighborhood's rooms: it is almost as though the diamonds, and not just their merchants, are part of a holy endeavor.

Chapter 11

The Diamond Detectives

On Labor Day 1994, Dan Ribacoff's beeper went off. Dan, who runs International Investigative Group, was at his home on Long Island, enjoying the holiday. The call was from an insurance adjuster at Lloyd's of London, Tiffany & Co.'s insurance company. When he got on the phone, the adjuster said to him, "Dan, I need you to come into the city. We have to do interviews. Tiffany's was robbed."

The night before, September 4, Tiffany on Fifth Avenue had lost $1.9 million worth of jewelry. In all 157 years of its existence, the store had never been so profitably raided. A security guard who had been on shift at the time of the burglary said that a man had followed him to the store, put a .357 Magnum up to his ribs, and forced him to lie to the other guards on duty. He was to tell them that the thief was his cousin and needed to use the restroom. Inside the store, the thief tied the security guards up with duct tape. He came away with about 420 rings, necklaces, earrings, bracelets, and watches.

"Well, we interviewed the guys who were there and the

guy who said he was accosted and his story didn't add up," Dan Ribacoff told me.

A second person on the case was Dan's older brother, Elie, the owner and manager of a security company. While Dan talked to witnesses, Elie and the NYPD detectives studied Tiffany's system for clues on how it was breached. It seemed to have been an inside job.

Dan and his men teamed up with the NYPD and divided the suspects, which included the security guards, into two groups. The detectives would shadow one group, the NYPD would shadow the second. They followed the guards, who were in their late teens and early twenties, "morning, noon, and night." They trailed them to the projects in which they lived, to work, and back home again. They caught some of the guards moving Tiffany jewels on the streets of Harlem. Then they arrested them.

But the prize suspect, the man who had pulled the gun, still possessed the bulk of the merchandise. His name was Mark Klass. "And Mark Klass," Dan told me, "was in the wind, as they say. We couldn't find him, so we employed our very good technique of putting a price on his head."

Dan helped things along by taking the reward poster to an individual with a stake in the matter: Mark's mother. He visited the housing project apartment that she shared with her son and asked her where he was.

"I don't know," she told him.

"Well, listen," said Dan. "You see this reward poster? It has his picture on it. We've put a twenty-five-thousand-dollar reward on his head, and downstairs my people are handing it out, but I don't really think they want the twenty-five thousand. I think they're probably gonna kill him to try to get the two million dollars' worth of jewelry."

That evening, Dan got a page on his beeper again. He was at the movies with his wife and some of their friends. Mark Klass and his lawyer were on their way to the Midtown North precinct, the police station that covers Tiffany (as well as Forty-seventh Street).

"Needless to say, Mark Klass came in with his tail between his legs, saying, 'Here's the jewelry. I'm sorry.'" When Dan told me this, he could barely get the words out because he was laughing so hard.

One of the first hits I got when I searched Dan Ribacoff's name on the Internet was a website for a professional bodyguard service that shows a picture of him wearing goggles, aiming a gun out of the passenger window of a car. The service is an affiliate of International Investigative Group. Like De Beers, IIG has many tentacles, but the most important is Worldwide Security Network, Elie's firm. The Ribacoff brothers work together on most cases. Between them, they've solved thousands of diamond and jewelry thefts, hundreds of them on Forty-seventh Street.

Physically, Elie and Dan Ribacoff are variations on a theme. They have similarly dark almond-shaped eyes, and both of their foreheads are striking—Elie's because it has a vertical indentation in its middle, and Dan's because it is expansive, straight and tall, like a house to which an extra level has been added. Both brothers know diamonds can make people ravenous and inspire them to do cruel and perilous things. They've made it their life's work to protect jewelers from the myriad dangers they face every day, because their family came to know these dangers firsthand.

In 1979, their father, a retired engineer, opened a small

jewelry operation on Forty-fifth Street. Elie was his partner and Dan helped out over the summer. Every day, they'd drive into Manhattan together from Long Island—Elie, Dan, and Mr. and Mrs. Ribacoff. Mrs. Ribacoff went to get the coffee, Elie parked the car, and Mr. Ribacoff opened up shop.

One Monday they drove in early. Mr. Ribacoff was going to meet a courier for a New Rochelle jewelry store he'd sold to before. The courier was supposed to collect merchandise for a traveling salesman who was in a rush to get on the road. Mr. Ribacoff unlocked his office and set up. The doorbell rang soon after. When he answered the door, there were two men standing outside. The courier had brought someone with him. They were armed with guns, a lead pipe, and an ice pick.

"Fortunately," Elie told me, "we had, I'll say, better security than most people, in that you could not get out if someone didn't buzz you out. You couldn't just leave the office, so they were both stuck in there. My father was unconscious on the floor. The people next door heard the ruckus. They called the building super. They thought that he was having a heart attack or something, so the building security and a couple of custodians came in and peeked through the bulletproof window."

When security arrived, both burglars ran for separate fire exits. The shop courier escaped; his accomplice was thwarted by a jammed door—fresh paint. He served twenty-five years in jail but wouldn't talk about his partner in crime, and they never found the other man. By the time Elie and Dan got there, paramedics were tending to their father.

"A good portion of his scalp was hanging off from being beaten with a lead pipe," Dan remembers. "His shirt was all bloody from being stabbed with an ice pick." Mr. Ribacoff survived the attack, but he is legally blind as a result of it.

Even if they are lucky enough to come out alive, the overwhelming majority of robbed jewelers go out of business, because by the time they are compensated by insurance—if they are compensated at all—they have already defaulted on their rent and employee payments. "You know, you work hard, you invest a lot of money, and then one day, either you lose your life or you lose your life savings," Elie lamented.

For the most part, the Ribacoff brothers talk about the burglary the way they talk about any other case, listing off the details of their father's assault with the cadences of professionals. But there was a short pause before Elie told me, "That whole incident was traumatic," a tinge of breathiness in the last word. In general, their voices don't belie trauma. It's their vocation that does. By 1995, Elie and Dan had left behind jewelry sales and switched full-time to the security and investigative businesses. It's as though they are still looking for the man who got away.

In the meantime, they've managed to hunt down countless other offenders, making Forty-seventh Street safer for jewelers. Elie, who took on the electronic aspect of the operation, has plastered the district with double-door "mantraps," panic buttons, retina scanners, and hidden cameras. He is responsible for the technology at the Diamond Dealers Club entrances that records people's fingerprints, faces, and ID cards. When I visited him in his small satellite office on Forty-seventh Street, a camera on the ceiling pointed my way.

As a teenager, Elie watched "all the cop shows on TV" and loved *Mission: Impossible.* He wanted to master the gadgets and hidden cameras he saw the characters using. On his eighteenth birthday, he joined a special task force of the NYPD as a volunteer, walking the beat in different New

York City neighborhoods, which proved just as useful. "You don't often see the sophisticated Hollywood types of crimes with Tom Cruise rappelling down on these wires. Most crimes are crimes of opportunity." All it takes is a few days of surveillance to get a sense of what time a jeweler comes to work, whether he opens and closes shop on his own or with someone, where he keeps his expensive goods, and when he is most defenseless. "It's very low-tech but very efficient."

Elie knows that gadgets alone do not make a jewelry office impenetrable. "A camera is not gonna stop a bullet. Bullet-resistant glass is not gonna keep someone from robbing you when you're outside of that cage or that partition." In fact, just about a year after I worked for my father, thieves shattered one of Forty-seventh Street's exchange windows in broad daylight, stealing gemstones by the handful.

Usually, there aren't more than two million-dollar-plus thefts on the street in a year. But this does not account for jewelers who choose not to report thefts they've endured, because citing the crime would out them as carriers of undeclared or even legitimate but valuable jewelry. (The latter scenario could potentially lead to another robbery.) Elie guesses that there are as many unreported thefts a year as there are reported. The majority of crimes are preventable.

After our meeting, I passed along some of the tips I learned from Elie to my father. Don't travel with merchandise. Vary your routine. Buy a cop on the street a cup of coffee and ask him to help you open up shop one day, just to be seen with a police officer. It was the first time I'd given my father tips on the diamond life, and I tried to remember all that Elie had told me about keeping jewelers safe. I didn't want to miss a thing.

While Elie cultivated a business devising safety schemes for jewelers, Dan and his company followed diamond thieves to the edge of the world—in one case, they caught a man in Panama who had robbed a diamond company in China.

Dan took on fake identities to track down gemstones. "Sometimes I'll pretend to be super-rich, and put on my gold Rolex with diamonds and take the Mercedes and pretend to be the hootin' falutin guy, so that they'll whip out the goods we're looking for if they're very high end. Sometimes it's where I'll go in and basically try to sell stuff to them to see if they'll ask me for ID and properly report the transaction, so sometimes I'll have to dress down and not shave for a couple of days and pour a little beer on my clothes to make myself look like a less-than-savory character." He's waltzed romantically into suspect jewelry stores holding hands with a female partner to shop for engagement rings, worn a hard hat to play construction worker, and camouflaged a surveillance van as a plumbing company car to film a group of burglars who rented an office on Forty-seventh Street. Once, he put on sunglasses and sold a bunch of pencils on a street corner while pretending to be a blind man.

I asked Dan where he picked up his talent for acting.

"I was always a great actor and always in the school play." His talent has put him in some risky situations. Once, he entered the home of a Gambino crime family member—"very high member, as a matter of fact," Dan said, his voice breaking into a nervous, high-pitched laugh. He pretended to be a buyer on the lookout for stolen jewelry and was wearing wires to collect evidence for intent. The NYPD waited outside. If he got into trouble, he was supposed to say, "How 'bout those Mets?"

"That was the—" Dan interrupted himself, laughing.

"That was the code word. 'Hey, how 'bout those Mets?'" he said loudly. He laughed again. "Bang, down comes the door, you know?" Thankfully, he never had to bring up the Mets.

Both Elie and Dan walk around the city armed. Because it is their job to catch criminals, they are the frequent target of resentment. Dan confronts this with cockiness. When a convict he had turned in asked him, "Do you always carry your gun?" Dan replied, "I carry it three days a week, but you guess which three." The man had glimpsed Dan's firearm at the station where he'd brought him. "I took my jacket off. It was hot, and I had it on my hip, and he decided he would be a cutie, you know, try to intimidate me, so I certainly don't back down, and I think he was a little shocked and taken aback. But I've had people ask me, 'Have you ever shot anyone?' and the answer is 'No one who hasn't deserved it.' That's the standard reply."

The criminals in Dan's and Elie's diamond cases aren't always obvious perpetrators. A diamond dealer declaring a loss is often a suspect himself. Especially in difficult times, jewelers find creative ways of committing insurance fraud. One man shot himself in the arm inside his own office. "Oy! Oy! I've been shot!" Elie took on an old man's Yiddish accent to imitate the 911 call. After an analysis of the bullets, the office wall, and the jeweler's gash, Dan and the specialists he was working with concluded that the man was lying.

Other times, all a case requires is common sense. A Forty-seventh Street jeweler gave himself away when he told the police that two men raided his office, but forgot to take off his gold Rolex watch, an item thieves would surely have stolen.

When I asked Dan what is the most surprising thing about Forty-seventh Street, he said, "The amount of stolen property that goes through there. It really is unbelievable." He estimates that half of all stolen jewelry makes its way into the trade. Though there hasn't been a homicide associated with a diamond district theft in years, there is enough burglary for Dan—who solves non-diamond crimes as well, and is so respected in his field that the United States hired him as an investigator in the 1998 African embassy bombings case—to make a triweekly appearance on the street.

It doesn't take a Mafioso to steal a diamond. Thieving techniques are as diverse as the gemstones themselves. One option is switching stones: the offender calls a supplier, asking to see a particular type of stone, say three-carat round diamonds of high quality. He brings with him a diamond, or a piece of glass or cubic zirconium (a popular diamond mimic) of the same or similar size and shape but of inferior quality. As he's pretending to examine the stones he's being shown, he replaces one of them with his lesser diamond, walking away with an upgrade. "Some of these guys are really good magicians," said Elie. "I mean, they can really make things disappear."

But thieves are not always so thoughtful as to leave proxies. Sometimes they make off with hundreds of thousands, even millions, of dollars in merchandise. To pull off a robbery of such proportions, they trade on trust.

Most diamond dealers purchase their goods on credit, so a man's reputation means everything. Forty-seventh Street and the international diamond community form a vast grapevine where a seller can reassure himself of a buyer's dependability through questions. Does she pay on time? Would you give him a month to pay a hundred thousand dollars? How long have they been in business? How long have *you* been doing business with them? When was the last time you sold to this dealer? With whom do they deal? Where do they get their diamonds and jewelry? Sometimes a dealer will inspect the reference's references.

But even this intricate system can be overcome. The thief starts out small, purchasing inexpensive goods and paying for them in full and on time. He does this for a while, then increases his credit and makes good on it, often buying from multiple dealers in town to establish references. Then, once his credit has reached a high value, he'll collect as many goods as he can on consignment or with terms of delayed payment. Since by this time the thief's credit, whether real or feigned, is immaculate, sellers feel comfortable giving him time on reimbursement: maybe thirty days, maybe sixty days. It doesn't make a difference. They will never see the money. The buyer will either disappear or just not settle up. "It's not uncommon to see someone who has 'taken you' for thousands of dollars in an elevator with you," I learned from Elie. Bankruptcy and other complex legal situations protect a man who has failed to pay his seller.

Not all defaulted diamond deals are planned. A dealer may be in over his head with debt and, as a last resort, purchase a large diamond on credit and try, without success, to save his business by reselling the stone. Or he may buy a stone when times are good, and the economy will dip before he has a chance to sell it. Now he's sold it for only seventy percent of its original value and owes the seller tens of thousands of dollars he simply does not have. In cases of debt, the money is gone, even if the man himself has not fled. "I've seen people that owe me money that they cannot pay," Elvis, the singing jeweler, told me angrily. "They take goods and they don't pay. And there is nothing you can do about it. There is *nothing.* If a guy takes your goods, he doesn't wanna pay you, you cannot do *nothing* about it." He broke into Bukharan, one of the several languages he speaks, then repeated, "Nothing!"

An arrest will only sometimes bring back the cash. Efraim Reiss got burned during one of his first business deals. He was working for his father when he paid a call to a respected office in the district. A man he'd never met came in to show him some rough. Efraim was interested. The office holder advised him to take it to his father, a rough expert, to determine its worth. The owner of the rough demanded collateral, so Efraim left thirty thousand dollars in diamonds with the man. They planned to meet back in an hour.

When Jack Reiss looked at the goods, he pronounced them fakes. "So we call back," Efraim said. "He says he's coming, he's in a restaurant, says he's eating." Jack had a feeling they'd been swindled. Efraim called the police, who spoke to the owner of the fake rough on the phone and threatened to detain him if he didn't return the goods. The man didn't show, but it was Friday afternoon and Shabbat was approaching, so Efraim had to make his way back to Borough Park.

A week or two passed. The man's phone was disconnected. The owner of an exchange where the thief had been spotted before the incident let Efraim use his surveillance system to print a picture. They showed the image to the exchange's security guards and asked them to keep a lookout. A few weeks after the original theft, Efraim got a call. The guards had seen the man in the exchange. Efraim called his detective, who spoke to two policemen on duty.

The thief was arrested outside a restaurant on Forty-seventh Street. Efraim spied the scene from down the street. The thief tried to argue—he would pay Efraim right away, just let him talk to Efraim—but they took him to jail. After the indictment, Efraim agreed to drop charges if the man returned the money and enrolled in a rehab program for the drug problem they'd discovered he had.

"That was my lesson for the first time," said Efraim, "and since then, I only deal with good people. Only reputable people, and since then, thank God, I did not have any problems."

Another way to steal a diamond is the bust-out, in which a thief amasses other people's goods and then pretends to get robbed himself. In 2007, an Israeli dealer, along with his brother, father, and another man, were said to have taken $3.3 million worth of diamonds they acquired on consignment from at least forty-five dealers on Forty-seventh Street, then blamed a fake robbery to account for one million dollars of the loss.

Sometimes merchandise goes missing without a clear suspect. I first heard about the 10 West diamond from Ken Kahn, the owner of that exchange. It's not uncommon for a stone to disappear from a booth. Tenants might forget to clear a piece of jewelry out of their display case at the end of the day. Or a diamond will spring out from between the rods of tweezers. "Diamonds pop," Ken told me, "they jump. They're very jumpy. Like a jumping bean."

In July 2008, a young diamond broker was peddling a 7.5-carat $750,000 diamond on behalf of a private customer. He entered 10 West with an envelope containing the diamond in his interior jacket pocket. He offered the stone to a booth holder, who turned it down. So he moved on to another building. When he got to the next customer, he checked his pocket. The diamond wasn't there.

The broker ran back to the exchange. He crawled on the floor, searching for the stone. He asked the booth holders if they'd seen it. "He's ready to kill himself, 'cause he's responsible. How are you gonna be responsible for a seven-hundred-thousand-dollar diamond?" Ken and the broker watched the surveillance tapes several times, but they didn't find any substantial clues.

The next character to enter Ken's story is the insurance adjuster, a friend and colleague of the Ribacoffs and a familiar face among the landlords and security experts of Forty-seventh Street. He is often called to the district to investigate missing diamonds. Months later, I would meet this adjuster, John Enders, for five minutes by the clock in the middle of Grand Central Station, before he had to bolt off to work on another case.

The night of the loss, Enders took home a few tapes of 10 West's exterior. He figured he could narrow his suspects down to those people who had left the exchange after the broker walked out. One man in particular caught Enders's eye. He seemed to be sliding an envelope into his jacket. Enders figured this character might have seen the envelope on the ground and taken it. He asked for more tapes, including interior coverage, to solidify his theory and spent thirty hours watching them. Now he had a clear idea of what had happened. The man he'd suspected had been in the back of the exchange at another booth and was on his way out when he noticed the envelope on the floor, and took it.

After corroborating the account with the security company that runs the surveillance at 10 West, Enders moved on to recover the stone. He got a tip (which he couldn't reveal to me) that led him to locate the man who'd taken the diamond—a booth holder at another exchange.

At first, when Enders confronted him, the jeweler denied everything, but after Enders threatened to go to the police, he admitted to having stashed the diamond in a safe-deposit box. Enders made what he calls "a gentleman's agreement" with the jeweler: Lloyd's of London, the insurance company Enders worked for, would not actively seek prosecution. The jeweler, in turn, would meet Enders and the broker at an office in the district with the stone. The next day, everything worked according to plan.

I asked Enders why he trusted the jeweler to show up.

"Well, he said he didn't have the diamond with him. He said he had it in a safe-deposit box somewhere, and I guess he wanted to talk to a lawyer or something. I could completely understand that, and I knew who he was, I knew where he worked, I knew he wasn't going anywhere. He had a business to run. The last thing he needed was bad publicity on the streets of New York in the diamond district." And so the rules of trust and reputation played a role in bringing back the missing stone.

I hadn't known that there were so many ways of stealing diamonds. But when I met with David Abraham in the Diamond Dealers Club, he told me simply, "Nobody has to pickpocket—although there are pickpockets—'cause all you gotta do is just find a way to buy merchandise and not pay for it. And there's plenty of situations like that."

To a large extent, Forty-seventh Street tries to govern itself. From the early days of its existence, the Diamond Dealers Club has acted as the district's moral authority. One of its most important hallmarks is its arbitration system, whose tribunals are a legendary way of solving the industry's disagreements from within.

If members are unable to resolve a conflict, their case

goes to a court of three. A date is set for the hearing, and a panel of five arbitrators is picked randomly from a hat. The five include two alternates who hear the case but don't judge unless a core member is unable to attend a hearing or decides to bow out of the case. All five arbitrators listen to the evidence and testimonies, but when it comes time to deliberate, only the core tribunal remains in the room. The ruling is not always pronounced immediately. It can take arbitrators hours or months to reach their decision. Sometimes they lose sleep over it, or expose themselves to the fury of the indicted.

One former arbitrator told me about a dispute that pitted a New York dealer against a foreign dealer. She and the other arbitrators in the case ruled against the local dealer. The man refused to talk to any of them again. Soon after the case, he fell ill. "I felt very very responsible," she confessed to me, "but when you're an arbitrator, you go according to the paperwork and what you can read and what you can see, and that's what we did." Still, she felt a burden in judging others and never pursued a second appointment.

Elvis, who is also an arbitrator, says he knows who is right and who is wrong immediately after hearing the facts and testimonies. "Let's say you bought something from me, and you're not paying me, who is right: you or me?"

"You," I answered.

"This is arbitration, basically," he said matter-of-factly.

I asked him if there weren't sometimes more complicated cases.

"Yeah, there's complicated, but if you really think about it, it should never *be* complicated."

Not everyone sees it this way. One evening David Abraham was leaving his office building with his wife. At the time, he was the Club's vice president and thus chairman of arbitration. Outside the building, a man whom the court had ruled against came up to him and ordered him to annul the decision. David would not. They argued.

"Believe me, you don't wanna have the IRS after you," threatened the dealer.

"It felt like a sudden attempted hijack to my heart and brain and mental peace of mind," David remembers. The IRS poses a menace to even the most scrupulous of diamond dealers, because once word gets around the street that the agency is interested in his records, other dealers will think twice before trading with him. But David had learned to remain composed. He calmly told the Club member to do whatever he wanted, took his wife's arm, and went home. The IRS never came after him.

Arbitration can be emotional, not to mention expensive, costing a minimum of two hundred and fifty dollars and varying by the worth of the diamonds or money in dispute. It can destroy friendships. And if a person does not abide by the rulings of the court, his picture will hang on the Club bulletin boards, advertising his insurgence to the entire community.

"I always try to work it out verbally between the people," David told me, "before anything comes written, before any bad blood is spilled."

If members from other cities' diamond clubs are involved in a dispute, the New York Diamond Dealers Club will contact the World Federation of Diamond Bourses, because "then you have a jurisdictional issue. Where should the arbitration be held? Where was the business done? Where was *Mazal* made?"

Arbitration is such a powerful judicial scheme that New York's state courts maintain its verdicts in the vast majority of cases, even when a mistake has been made by an arbitrator. The courts are so supportive of a business dealing with its own clashes that, most of the time, they are only willing to override a decision if there's been a faulty procedure or a fundamental problem with one of the arbitrators, such as bias or fraud.

Still, despite the diamond court's prominence, some feel it is limited in its power to ease dealers' financial risks. David thinks part of the problem comes from the stringencies of the U.S. legal system. "The Diamond Dealers Club is quite a different organization from the diamond club in Israel or the diamond club in Antwerp, not just because it's a very different country and each country has their own culture, but this is a very litigious society and the threat of lawsuits is always here. And it doesn't allow the DDC to be able to take the quick action that they sometimes need to, to stop a person that maybe is doing a criminal activity like cheating people." It is for this reason that one man decided to take matters into his own hands.

Out of a modest office on Forty-seventh Street, Zev Oster, a fourth-generation diamond dealer, runs something of a diamond vigilante group. Like the criminals he is trying to thwart, Zev doesn't use baseball bats or guns. He uses the Internet. Jewelersalert.com is a website where members can broadcast bad experiences they've had with companies and learn the names of those who have been known to delay payment or have gone bankrupt.

"If you ask anybody how much money he lost in the industry since he's in business, he'll just laugh at you," Zev said humorlessly, "'cause it's outrageous what happens. And you see people going out of business every day because it happens."

When I met him, Zev didn't strike me as an avenger. I saw a slender man of thirty-two with a pale and youthful face, the face of a boy who might turn into a Torah scholar someday. He wore half-rimmed glasses and a black velvet *kippah* on his head, and tucked his side curls behind his ears, where they were still visible but not prominent.

Zev came up with Jewelersalert.com in 2007 after his family business, American Fancies, which he shares with his father and brother, became the victim of a scam. The company mailed a $25,000 diamond to a customer in Chicago. They had checked the man's credit record, found nothing suspicious, and sent off the stone. But when Zev contacted the dealer for compensation, the man had disappeared. It was Zev who had found this client, and so he felt responsible for leading American Fancies into a bad deal. "Obviously, in the diamond industry twenty-five thousand dollars isn't a tremendous amount of money, but it's twenty-five thousand dollars, right?"

After his loss, Zev called the man's references and asked them, "'Hey, how come you said this guy was good?'

"'Oh, oh, that guy? He's not good anymore.'"

Zev produced a quick, mournful tongue click, as we sat talking in his office. "So there was a point that he actually went sour, and people *knew* about it, but *I* didn't know about it."

He decided to do something. For weeks, he spent his nights holed up in his bedroom, gathering information. Then he launched his site.

I took a look at Jewelersalert.com. It is organized in forum topics: Past due, Slow payers, Schleppers; Scammers, Thieves, Fraud; and Resolved, where people post notifications when a situation has been righted. The threat of a blemish on a jeweler's permanent record seems to motivate them to stay on top of payments. Numerous dealers have told Zev that after voicing their complaint of an unpaid bill on Jewelers Alert, they were finally compensated. Because the diamond business functions on reputation, the muscle behind Jewelers Alert is powerful. Sometimes hundreds of people log on to the site at once with names that sound like high school instant-message aliases: RingFinger, dboy, diamdealer, and kingdiamonds.

There are a few exceptions to the permanence rule. In one case, a false rumor was posted to the website and nearly ruined a dealer, making it all but impossible for him to receive goods on credit. Zev took down the post and suggested that the dealer issue a counterpost, but the man decided it would be best to wait out the disaster rather than engage.

In general, Zev is a cautious modernizer. To guard against liability, Jewelersalert.com uses memos, invoices, and FedEx tracking numbers to verify complaints, and Zev can sue anyone who posts libelous material on his site.

When it comes to Jewish libel—*Lashon Hara*—Zev is not worried. In fact, he sees the preemptive job that Jewelers Alert does as a duty, a mitzvah. The website's motto is "Jewelers Protecting Jewelers."

I remembered learning all of the intricate rules that governed what Jews were and were not allowed to say about other people. Zev's sideburns made me think that he did, too. I asked him if he consulted with his rabbi.

"I did speak to my rabbi, but . . ." He laughed. "It was gonna be done either way. It had to be done. I'm sure when you go up to God and say, 'Listen, you know, this guy's ripping off your money. Am I gonna get punished for that?' He'll say, 'Yeah, I don't think so.'"

As of yet, jewelersalert.com does not pay Zev's bills. But even if it does in the future, Zev does not plan on leaving American Fancies. The site more or less runs itself, so when he's not busting diamond crime, Zev is an average Forty-seventh Streeter. He sits at a desk in the back room of a small office. Every morning, he comes in on the Monsey bus, where he prays with the other riders. He himself is religious but not Hasidic. (His brother is.) And, like many other dealers, his entry into the business was more or less an inevitability.

Six months after his wedding, Zev was in Israel, studying the Torah and Talmud, when his father called him up and said, "'Okay, you're taking two courses. You're taking a rough course and a diamond course.' And that was the first time I started learning about diamonds."

Zev is no idealist. He believes that the *Mazal* code has withered over the years. Back then, "you said *Mazal*, it's your stone. You eat it. You wouldn't return it." But today, customers will often try to give back a diamond after agreeing to buy it, and sellers will often accept it to keep their clients happy. Because nonpayment is not a new problem, diamond people came up with ad hoc efforts to exchange information before Jewelers Alert existed, like posting notes to the Diamond Dealers Club bulletin boards. But Zev maintains that this does not really constitute a successful alert forum. Like Martin Rapaport, he went beyond the framework of the contemporary diamond trade to fill a need he recognized. When I ventured to point out similarities between the two, Zev answered reverentially, "I don't compare myself. He's huge. He's my role model."

The thefts Zev deals with every day are the ones that don't involve heists or gunshots, only a violated *Mazal.* The other kind, the dangerous kind, is comparatively rare in the district—only one or two armed in a typical year. According to the Jewelers' Security Alliance, burglaries, theft, and off-premises crimes in the United States all decreased in 2009, both in value and in frequency. Though a total of $45.9 million in jewelry was stolen in the country, average losses were down from previous years. The figure, which sounds like a lot, was put into perspective for me when I learned from Elie Ribacoff that, on Forty-seventh Street alone, millions of dollars' worth of gems and jewels are carried around every day. Although the street is teeming with diamonds—more than two thousand jewelry-affiliated operations populate the neighborhood—it is still a relatively safe place. There was a time when it wasn't.

The diamond district my father joined in 1984 was a jungle. I think of my parents' integration into America as something like the cutting of a diamond: a process of refinement. But back then, when they first moved, the diamond district was not refined, and in some ways, neither were my parents, at least not for the life that awaited them in New York. On Forty-seventh Street, thieves patrolled the sidewalks. Industry insiders would aid the robbers by identifying dealers on the street. These were called "finger jobs." Then, in the late eighties, the Colombian gangs moved in. They hung around corners in groups of three to five—a spotter or two and a few accomplices—fixed among the waves of moving pedestrians. They put ketchup on people's jackets and pointed out in horror that they were bleeding, or poured coffee on them and grabbed their briefcases. They dropped money, a hundred-dollar bill laid precisely atop a roll of one-dollar bills, then stopped people to ask them if the money was theirs, while an accomplice snatched their goods and handed them off to a third conspirator, who disappeared down the block. Once, my father saw a thief approach a man who was getting into a taxi and point at a bunch of bills on the ground near the car. The man placed his bag into the cab before bending down to get the money, and a second clip artist came around from the other side of the car to swipe it. When the man realized what had happened, he called for help, but the bandits had already fled.

My father's friend Lester was leaving the district one day, walking eastward, when a man stopped him and said, "Somebody put ketchup on your jacket." Knowing the trick, all Lester said was, "Okay." He continued on his way. Some paces later, another man approached him.

"Hey, there's ketchup on your back." Lester's theory is that they wanted him to put his briefcase down on the ground so they could grab it as he checked his jacket. He kept walking, and the conmen kept stopping him. There must have been ten of them. On Park Avenue, another one, this time a woman, asked him if he'd dropped money on the ground. "And then I really started getting afraid, and I said"—Lester paused, breathed in, and shouted to me on the phone—"'POLICE!' and everybody scattered." Then he went back to his hotel to change his ketchup-stained jacket.

My father tried to steer clear of the traps thieves laid for diamond men by keeping one step ahead of them and learning from the misfortunes of others. So when a man dropped a bundle of cash at his feet while he himself was getting into a taxi and said, "Sir, you dropped some money," my father summoned his most threatening demeanor and came back with "It must be yours, and if you don't get the fuck away from me, I'm calling the cops." He leapt into the cab and ordered the driver to go.

Sometimes dealers fought back. One man tripped a pick-pocket on Forty-seventh Street. A group held the thief down until the police arrived. But what finally reined in the crime was a series of meetings between representatives of the diamond industry and the New York City Police Department, during the nineties. Eli Haas, who would go on to become president of the Diamond Dealers Club, attended these meetings. He told me that, at first, the NYPD responded to the Club's pleas for increased protection by saying that they couldn't focus on Forty-seventh Street more than any of the other thousands of city streets. Finally, Haas and the Club leadership managed to convince the police commissioner that Forty-seventh Street, a place where men and women walked around with diamonds in their pockets every day, was simply not like every other street.

And it wasn't. In the mid- to late eighties, diamond men were being pursued in their offices, factories, and shops. Burglaries were commonplace. A dealer was beaten and robbed of $1.5 million in jewels by two thieves with guns. A man and a woman who worked for a wholesale jewelry company were shot to death during a midday break-in. A jeweler was murdered in his office foyer, and another was found through the glass door of his office by an employee of the building—shot dead on the floor.

My father felt the danger. He was leaving his building one night. It was late, and the few streetlamps on Forty-seventh Street provided little light; this was long before the massive diamond columns were erected. Two men were waiting outside his building. They came toward him. My father kept walking, but they followed. As he started to run, they picked up their pace. His heart beat furiously. They were younger than he was, and he knew they were faster.

Down the street, my father spotted two security men, employees of the same alarm company he used. He approached the guards. He didn't shout for help, he didn't beg. He laid out his request in business terms, explaining to them that he was a client of their company's, and asked if they would please walk him to his car.

The security guards walked with him, one on either side, and behind them, the two men who had threatened him followed, like a procession. Boosted by the assurance of two guards beside him, my father put his hand in his chest pocket, as though holding a gun, and said to the men behind him, "You don't wanna follow me."

They were unimpressed. "You ain't got nothing," they said. "Show me." They followed my father and the guards all the way to the car. In the garage, before my father drove away, the two men said to him, "We'll get you next time," like villains in a low-budget motion picture, though the danger was real.

Most perilous was life on the road. Nowadays, jewelers tend to ship their goods with armed movers for safety reasons and because many insurance companies have boosted their premiums for traveling salesmen. But back when my father was a newcomer to the business, coverage was not dependent on how many days a dealer spent on the road or to which countries he flew to try to sell his goods. Many diamond men carried their merchandise on their bodies or in their luggage when they traveled, instead of paying companies such as Brink's or Malca-Amit. This made some business trips particularly dangerous. My father was one of these men, but two incidents made him change his mind.

The first scare happened in the late eighties, just after he landed at Miami Beach's airport late one evening for an antique jewelry show. It was show season and jewelers were pouring into the airport from all over the country. The car rental center was situated in the shadowy outskirts of the airport. When my father finally got behind the wheel, he noticed a car with dark windows parked on the shoulder of the road. As he began to drive, the car hovered alongside his window. My father sped up. The other car sped up, too. Then a second car pulled up on his right. "That's when I just hit the gas pedal, just hit the gas pedal all the way. I floored it basically." But the cars kept trailing him, "and then I knew that I wasn't just hallucinating."

My father drove like a madman, flouting rights-of-way and red lights, hoping to draw a cop's attention. Finally, "I decided to just make a left at some point and I just braked a tiny bit in the last moment, just pulled the steering wheel over to the left, and, like in the movies, burning, smoking tires." The rental tore around the corner; the two cars hadn't been able to keep up with my father's surprise turn and rode straight into the distance.

When he'd lost the cars, he coasted along a large avenue. "I just kept speeding and hoping that cops would stop me, and then I would tell them what happened, but you know when you *want* a cop to stop you, there're none around. So I just kept going all the way till I hit Miami Beach."

He reached his hotel and found his dealer friends at the bar. "Lester was there and a whole bunch of other guys were there and they saw that I was white as a bedsheet. They asked me what happened." That night, instead of the cops, it was the diamond men my father confided in.

The fact that jewelers are in a dangerous business brings many of them closer together. My father met Lester at an auction in Canada in the early eighties, and they have been friends ever since. Though Lester lives in Los Angeles and my father in New York, they talk every week, mostly in English, though in moments of intimacy or humor, they sometimes turn to Yiddish. Lester and my father have owned jewelry together, celebrated family occasions together—they have even run from thieves together.

A few years after the Miami incident, on a trip to a jewelry show in Las Vegas, they were returning to their hotel one night from Harrah's, where they'd been displaying their goods. They rode in a limo—one of the perks Lester had access to because he was a high roller. The other perk was a duplex hotel room he was sharing with my father. Perhaps their pursuers had seen them in the fancy car and thought they were loaded with cash or goods. At the hotel, a small group waited for the elevator with them. They all got in, but no one pushed a button. My father had a bad feeling. He asked one of them where they were going and pressed that button and his own. Then he said to Lester, in Yiddish, "It's not kosher." At first, Lester didn't understand. "The group, it's not kosher," my father said again, and watched Lester's face change suddenly. The floor the men had requested came and went with no one getting out, so my father, alarmed, pushed the next button he could, the bottom floor of the duplex. He said to Lester, "We're getting out now."

When the elevator doors opened, Lester and my father ran to their hotel room and slammed the door. They called hotel security, who said they'd heard of other suspicious behavior in the building but that it would be futile to try to find the men, since they had almost certainly fled.

My father was only a few years older than I am now, a relative newcomer to America and the district. He'd always known that there would be close calls, but this was the livelihood he had chosen. He shared these stories with my mother but withheld them from me.

It took a long time for him to let me in on this part of his life. One day, after I'd already stopped working for him, we were in the office, talking about break-ins.

"Personally," he said, "I think I would fight."

"If they have a gun?"

"Well, they have to get close to you at some point."

"Not with a gun," I said.

He said the thieves would want to tie him up. Then he found a pair of sharp scissors on his desk, held them pointed side up, and smiled mischievously.

I was used to my father's preparation for worst-case scenarios. His fear and secrecy were fixtures in my life before he told me about the dangers of his business. My family absorbed them. We were never exempted from our responsibility to keep quiet, even at home, even regarding the most mundane parts of our lives. My family lived in a co-op apartment building where private washing machines were forbidden, but it was a luxury my parents were unwilling to live without. So one day, they snuck a midsized washer-dryer up through the service elevator. For years they lived in fear of the superintendent, of the board, of being caught. To make matters worse, the super's office was located right near the laundry room. Surely he would notice our absence in the daytime traffic of laundry doers. So, to preempt suspicion, my father sent members of my family down on a monthly pilgrimage to the basement with a load of whites and colors. This was how we kept up appearances. But just in case, we locked the closet to the washer-dryer unit every time a maintenance man visited, as though the diamonds were in there, too.

Knowing what I know now, his watchfulness makes sense to me. I've noticed it among other dealers his age. If you press hard enough, many of them have stories of burglaries and thieves, of occasions when their ware put them at risk. And at times like those, Dan and Elie Ribacoff can do nothing for the diamond men. *Mazal* will not help. Arbitration will not help. They are on their own, away from the rules and constricts of the trade, carrying a load that some people would kill for.

Chapter 12

Blood in the Land of Diamonds

The word "diamond" comes from the Greek word *adamas,* which, before it came to mean diamond itself, used to indicate, in broad terms, the hardest material in the world. There is an Indian legend that diamonds came from a king by the name of Bala. So great a warrior was he that not even the gods could conquer him in battle. To dominate him, they asked him to offer himself as a sacrifice, and he agreed. When they burned his body, his bones turned to diamond seeds instead of ashes. Supernatural forces dwelled within them. Serpents, gods, and spirits grabbed the seeds, but they accidentally let some drop, and the diamonds fell from the sky into the waters, the mountains, the forests.

For a time, people believed that diamonds had the ability to heal madness, counteract poison, and keep demons at bay. They could also bring ill fortune. A stone that included red marks on its interior, they said, would beget death. The diamond's beauty has always been symbolically tied up with both salvation and toxicity. In the 1990s, this connection manifested itself in the modern diamond trade.

There was almost an inverse relationship between the splendor of an African country's natural resources and the well-being of its people. Sierra Leone spent most of the nineteenth and twentieth centuries as a British colony. Following the elections of 1967, only six years after it gained its independence, a series of coups threatened the country's political stability. In 1978, about ten years after the All Peoples Congress came to power, the constitution was amended to ban all other political parties. It was from the tyranny of this government that the rebel group Revolutionary United Front (RUF) allegedly sought to free the people when they entered Sierra Leone from Liberia and launched their offensive in 1991. But early into their insurgency, the RUF captured the country's profitable Kono district diamond fields, and the focus of the war shifted from politics to precious gems.

The RUF terrorized the people of Sierra Leone for eleven years, funded by diamonds. The rebels put their captives to work in the mines, and diamonds allowed their soldiers to dress themselves in AK-47s. In parts of the country—whose currency was essentially worthless—the gemstones were used as cash. Diamonds were power. They determined who lived and who died; who was to be feared, revered, and obeyed.

To enlarge their army, the RUF kidnapped, drugged, and brainwashed thousands of children, turning them into boy soldiers. By the time the war ended in 2002, it had killed tens of thousands, displaced approximately a third of the country's populace, and mutilated some three thousand people. In Sierra Leone, people voted with their hands; amputations were a warning to those who dared to vote against the RUF.

The rebel group lined people up and forced each to place an arm on a tree trunk, then hacked off either their hand or entire arm, sometimes two arms. They played cruel games, filling bags with pictures of different body parts and making villagers reach into the bags to choose their fate. If the slip depicted an ear, the villager lost his ear; if it depicted a foot, the villager lost his foot. Following an amputation, the rebels sometimes coerced people into running to the nearest clinic. The movement would cause the victim's heart to pump blood faster, forcing it out of the stump at lethal rates. Sometimes the rebels would take bets among themselves as to the gender of a pregnant woman's unborn child. When the bets were in, they would cut open her stomach to determine the winners.

Ian Smillie, who served on the United Nations Security Council's expert panel on Sierra Leone and visited the country in 2000 and 2001, told me that, despite the frenzied character of RUF raids, "what they were doing did have a strategic purpose. It was to scare people away from the diamond areas, and from other areas when they were interested in foraging." The amputations were "a terror tactic, pure and simple, and if you knew the rebels, you didn't need to be told twice to get out of town."

This was no ideological revolution. Sierra Leone's war was not only made possible by diamonds, it was fought *for* diamonds. Diamonds were sold to acquire arms with which more diamonds were obtained. They were the means and ends of battle. First, Smillie explained to me, the gems were mined, often by forced labor. Then smugglers would take the stones to other countries such as Liberia, or at least to the border of Liberia. Liberians would then supply the RUF with weapons in exchange for the stones. The majority of

RUF men, especially the fighters themselves, did not acquire much wealth. "Most of them weren't in it for the money, at least not in the short run—they wanted power."

The fall of the Soviet Union had produced a multitude of available guns, many of them either authentic or imitation Kalashnikovs (AK-47s), which Smillie calls a "favorite weapon for rebel armies in Africa." Arms dealers would obtain the guns in the former Soviet countries, forge end-user certificates to places such as Burkina Faso and Togo, then use those countries as stopovers or even skip them altogether, sending the weapons on to their actual destinations, like Monrovia, the Liberian capital.

In its own way, Monrovia became a diamond city, shady brother to New York, Antwerp, and Tel Aviv. Its stones would trickle into legitimate diamond centers. Greg Campbell, author of *Blood Diamonds,* documented how merchants would travel to Liberia, purchase the goods, and export them into Belgium. Among the RUF's customers, says Campbell, was the terrorist group Hezbollah.

Of course, diamonds that appear to be legal are more salable than blatantly illicit diamonds, so the vendors would forge invoices with Liberia, Guinea, Gambia, and others as the countries of origin. The path of diamonds from Sierra Leone to Belgium is supported by an embarrassing discrepancy in numbers. Belgium was listed as importing billions of dollars in diamonds from Liberia during the nineties. "Now, Liberia doesn't have billions of dollars' worth of diamonds," Smillie told me. "The best they could ever do in a year is maybe ten million at the outside." Similarly, Guinea's exports between 1993 and 1997 were recorded at only 2.6 million carats, while Belgium reported admitting 4.8 million carats from Guinea.

In 1996, I received my first serious piece of jewelry, a yellow diamond necklace, on the occasion of my bat mitzvah. I have no idea where it came from or who mined it, who bought it and sold it, or the countries through which it traveled before I strung it around my neck, a too-fancy ornament for a twelve-year-old girl.

Sierra Leone was not the only country plagued by diamond-driven conflicts. There was the Ivory Coast, the Democratic Republic of Congo, the Republic of Congo, and Angola, whose rebel forces called themselves UNITA.

In 1998, Global Witness, a nongovernmental organization (NGO), issued a pamphlet called *A Rough Trade: The Role of Companies and Governments in the Angolan Conflict,* focusing on that country's civil war. On the opening page, a picture of a hand filled with rough diamonds melts into a photograph of a soldier draped in loops of ammunition and holding a machine gun.

A Rough Trade said that dirty diamonds were trickling into the mainstream market, that smuggling from Angola persisted despite an embargo the UN had placed on unofficial (uncertified) Angolan diamonds earlier that year. Purchasers the world over were buying stones that had been sold by rebel vendors for the purpose of their war. Global Witness pronounced diamonds the facilitators of this conflict. They put a lot of the burden on De Beers, as the bigwig of the industry, and Belgium, as the trade's geographical center. Among their charges was that De Beers purchased diamonds in Angola during a time when most of the country's supply lay in the hands of UNITA rebels. Supporting information included excerpts from De Beers' own annual reports. One such report from 1996 read: ". . . the increasing outflow of Angolan diamonds to the major cutting centers, much of which De Beers was able to purchase through its outside buying offices."

Global Witness recommended that the UN run special inspections on South Africa and the UK, in addition to Belgium and Israel, since these were countries where sizable numbers of unofficial diamonds were purchased. But it was not just countries they held responsible. Their analysis of the conflict diamond problem went to the heart of the diamond trade's structure, some of its most long-standing features: its opacity, its virtual immunity to outside regulation, and the network of middlemen who participate in the diamond chain. "The diamond industry does not have to deal with public scrutiny in the way that many multinationals now have to," they wrote in the 1998 report.

Soon, all that would change. The next year, Global Witness and three other NGOs unveiled a campaign called Fatal Transactions. They dispatched packets of information, including death counts from the wars in Africa, to important newspaper editors. In 2000, Partnership Africa Canada (PAC), the NGO where Smillie worked, published a report entitled "The Heart of the Matter," which publicized the links between Sierra Leone's warfare and those beautiful gemstones the country happened to produce.

As the blood diamond story unfolded, the diamond industry and those who relied upon it feared a boycott was on its way. It caused enough alarm that in November of 1999, Nelson Mandela himself spoke out against across-the-board diamond sanctions: "If there is a boycott of diamonds," he warned, "the economies of especially two countries, Namibia and Botswana, will collapse and we want to avoid that."

De Beers hurried. By early 2000, they had stopped acquiring diamonds from Angola as well as the open market. From then on, the only African diamonds they would buy were those that came directly from their own mines. There were also several personalities within the industry who played a role in the fight against blood diamonds. One of them was Martin Rapaport.

In 1999, Charmian Gooch, the cofounder of Global Witness, paid Rapaport a visit. "So," he told me in his Vegas hotel room, through the Pringle chips he had just put in his mouth, "she called me up, she said, 'All right, I'm here. I wanna talk about diamonds, but no one wants to meet me.' I said, 'No one wants to meet you? Come to me.' And she told me what was going on." Through Gooch, Rapaport learned about the war that was tearing Sierra Leone apart.

"I believed her." He also believed that diamonds could be the solution to the problems they had caused. But he disagreed with Gooch on how to go about exacting change. Gooch wanted to track the flow of illicit diamonds from Sierra Leone. Rapaport thought this was an unrealistic goal. He turned to food for a model.

"I introduced the concept of kosher diamonds to her," he told me, still chewing, "and we started to think about how we could stop the war in Sierra Leone, not based on following the bad stuff but instead, follow the good stuff, and I just took the laws of kosher and thought about how we could apply it to the diamond industry." He recommended positively reinforcing the trade in legitimate diamonds by offering them a market. He believed that if certifiably clean diamonds were available, most people wouldn't buy dirty ones.

The year after Rapaport met with Gooch, he flew to the still warring Sierra Leone to meet with Foday Sankoh, the head of the RUF. A UN World Food Programme helicopter carried him into the country. When he came face-to-face with the rebel leader, Rapaport used the skill that every diamond man knows best: negotiation.

How does one even negotiate with a rebel leader? I asked Rapaport.

Casually, he explained. "You say to the guy, 'Listen, if you come out of the area and you become a good guy, you can still do diamond business, you can do whatever it is, as long as you become a good guy. Leave, please, get your people out of these diamond areas. It's creating terrible problems for everybody in the diamond industry in the world.' And it was also like, we're gonna make sure that no one buys your diamonds. But he kinda blew me off."

When Rapaport went to see an amputee camp, though, he knew he had to do something. The camp was filled with limbless people wandering around. One of them approached Rapaport with a baby, Maria. She was missing an arm, as was the man holding her. "Tell them what happened," the man said to Rapaport. He meant the rest of the world. "Tell them what happened." Rapaport thought of Auschwitz, where his parents had been imprisoned during the Holocaust.

"Tell the world what's happened here. That was my mission. That was my message. It was given to me by some guy holding a six-month-old baby."

After his trip to Sierra Leone, Rapaport—along with a few eminent diamond dealers—met with the U.S. State

Department to talk about what could be done. The dealers, he said, thought he was crazy when he presented the kosher diamond idea. They also suspected him of encouraging this monitoring plan in order to line his own pockets.

In April of 2000, Rapaport published an article in his magazine urging the diamond industry not to close their eyes to the suffering diamonds were causing. He called it "Guilt Trip," and in it he printed pictures of Sierra Leonean corpses. Many people were enraged with him for publicizing these nasty images. He told me that some of the trade's luminaries literally ordered him to stop.

Smillie considers Rapaport "the first to break ranks in the industry and say, 'Boy oh boy, this *does* have something to do with us, and we really have to get on top of this.'" It was Rapaport who coined the term "conflict diamonds."

In May 2000, he and Partnership Africa Canada attended a meeting in Kimberley, South Africa, that launched discussions on an international method of control for the import and export of diamonds. Global Witness was there, De Beers was there, and representatives of several governments, including South Africa, Namibia, Botswana, Belgium, and the United States were there. The ideas that emerged from this meeting would eventually materialize into what is known today as the Kimberley Process. But not without a lot of work.

In the summer of 2000, the World Diamond Congress met in Belgium to talk about the disaster that had become affiliated with their trade and how they would deal with it. Ironically, 2000 turned out to be a great year for sales, but people feared this could change. Moreover, the affiliation between diamonds and murder was disturbing to much of the diamond community. The meeting gave rise to the World Diamond Council, a group rallying industry, bank, and international representatives in an effort to fight blood diamonds—and, according to Campbell, "to handle PR and spin control for the conflict diamond issue."

In the meantime, the U.S. Congress—the parliament of the biggest pool of diamond customers in the world—was divided. In May 2000, Representative Tony Hall (D.-Ohio) had read Rapaport's "Guilt Trip" aloud when introducing his bill for clean diamonds, which mandated a proof of origin for all diamonds being imported into America. Alternatively, Senator Judd Gregg (R.-N.H.) proposed what was, according to Smillie, a more lenient bill, which won the support of the diamond industry. Hall, a sandy-haired man with a vaguely Kennedy-esque face, became the NGOs' Congressional poster child.

Hall is an intensely religious Christian who once fasted for twenty-two days after Congress voted to eliminate the House Select Committee on Hunger. He has been nominated for a Nobel Peace Prize three times and likes to take on international crises in which success is measured not necessarily by the cessation of conflict but by the relative reduction of suffering.

He is not a diamond expert. The only one he's ever bought is nestled inside the modest gold ring he purchased for the occasion of his proposal to his wife. And before 1998, when Global Witness wrote him a letter, he had never heard of a blood diamond.

Then he went to Sierra Leone, and what he saw haunts him still. "Gee, I just remember, I'll never forget this little girl that I was holding," he told me, "and she was the prettiest girl—I mean the *cutest* little girl. She must have been about four or five, and she just had an elbow. That's all she had, so from her elbow to her hand it was cut off, and I thought, 'How the heck can you do something like that?'"

Hall told me about people without legs and young women missing noses. "I mean, it's hard to imagine having your nose cut off. I mean, how do you *cut* somebody's nose off, and how do you hold them to cut their nose *off*? Geez." He released a loud exasperated breath. Then he was silent.

While Congress and industry leaders scrambled to find solutions, smaller dealers on Forty-seventh Street felt a variety of emotions, ranging from assaulted to confused. All of a sudden, a product they'd been proud of selling for years was under attack. Some dealers felt diamonds had been unfairly targeted. Sal Lipiner, who was chairman of the Diamond Dealers Club board during the nineties and a longtime member of the industry, complained that NGOs weren't nearly as critical about the conditions under which gold was mined.

"Why?" he asked me rhetorically. It's not that gold is any more innocent than diamonds in funding wars. Rather, "I don't think you want to tangle with McMoRan, and I don't think you want to tangle with Rio Tinto, both of whom could put De Beers in their vest pocket." (McMoRan and Rio Tinto are exploration companies.)

Michael Goldstein, an antiques dealer in the district, told me what went through his mind: "This is terrible, but this war has nothing to do with diamonds. If they're gonna kill each other, it's all about the power in the country. And if it was cocoa, that would be it. It has nothing to do with diamonds." He wasn't the only one who pointed to other commodities with nagging histories. One dealer named oil, rubber, and wood. Another compared it to clothing made in sweatshops.

A foreign dealer I met told me he used to buy diamonds on the open market near the mines in South Africa, then sell these diamonds on Forty-seventh Street during the nineties. I asked him how he knew he wasn't buying blood diamonds.

"Because we were buying from legitimate sources all the time, and the conflict countries, you couldn't go to them. I don't know, it was a war zone or something." It's possible his stones were clean. But it's not provable. Though combat was limited to the continent of Africa, the diamonds were being dispersed throughout the world. On the other hand, from a technical point of view, the man is absolutely right. Even Smillie will tell you that "there's no such thing as blood diamonds until the NGOs expose them."

"No one's gonna say, 'Aw, I'm in favor of what's going on there,'" Michael Goldstein said to me frankly, "but you know that diamonds come from very nasty places." The truth is I didn't really know this, at least not until the movie *Blood Diamond,* with Leonardo DiCaprio, came out in 2006, long after many of the diamond wars had ended. I didn't know much at all about where diamonds came from. Neither did millions of other people. Surprisingly, these people included some dealers as well. When I asked members of the community about conflict stones, many of them answered with some variation of "I only deal with reputable people." If a diamond came from someone you'd been doing business with for years, whose *father* you'd been doing business with for years, it seemed implausible that it would somehow be tainted. Few had been pushed to examine the sources of their gems more than one or two steps removed, and that didn't necessarily come from a desire to sweep the ugliness under the carpet.

The foreign dealer who'd bought in South Africa even told me that he wasn't worried when the *Blood Diamond* movie came out. "I think actually it made good publicity for diamonds, and you see, the fishing companies, they take fish from the sea, they kill the fish, whatever. They slaughter animals. People aren't stopping to eat meat or fish, you know?" He said the film showed only the bad side of diamonds, "but certain other countries, they're living off diamond. People are living off diamond."

But others were tormented when blood diamonds came to light, knowing that, somewhere, the gemstones were the cause of suffering. A lot of diamond people had been the victims of tyranny themselves. Mayer Herz, a New York dealer, worked closely with Tony Hall and his chief of staff when they were drafting their bill to come up with viable legislation. He confided in me, "If you're gonna ask me what got me involved in this? I'm a child of Holocaust survivors. My father and my mother were married before the war and they were in two different places for three years and they survived the war. They had three children together before the war, who went up in the gas chambers, and I grew up with this. I mean, we knew this, we here and there spoke about it. When it came to what's going on in Africa, I felt that it was my obligation, and if you brought it to anybody's attention in our community, he was very receptive to doing something about it."

Rapaport believes that the very nature of the diamond trade places the lion's share of dealers in the staunchly anti–blood diamond camp. The industry tends to weed out unethical people, because they are the same types who would switch stones or steal. Those dealers who survive, especially those who last for generations, are likely to maintain strict values, values that would be at odds with purchasing conflict stones.

Which is not to say business wasn't on their minds. The industry had already seen how badly the fur trade had been hurt by advocacy. David Abraham, the former Club VP, remembers a woman with a fur coat getting sprayed with paint right on Fifth Avenue. He himself was getting questions from private customers—"What does a blood diamond look like?"—and demands such as "Please don't show me blood diamonds."

How did he answer them?

"I told the truth: diamonds look the same, like people look the same. Choose your friends and family with thought, and buy your diamonds from good provenance."

Tony Hall didn't splash paint on anyone, but he did picket Tiffany & Co. and Cartier jewelry stores on Fifth Avenue. Along with his daughter, his chief of staff, and a handful of others, he called out things like "Diamonds are not a girl's best friend."

"I wasn't the diamond dealers' favorite person for a long time," he told me, laughing a bit. But he also visited Forty-seventh Street. He took a tour inside a diamond-cutting factory, he met with Mayer Herz and other dealers, and went up to the Diamond Dealers Club. He was astounded to see the little notices that advertised lost diamonds on the bulletin boards. "You're not tellin' me that people would actually *return* these diamonds," he asked the dealers, and they replied that people would, that people did it all the time. Hall was impressed.

The other diamond center the congressman visited was De Beers. Before it was clear that Hall and Global Witness's efforts would translate into worldwide legislation, the diamond company tried to coax the congressman back into silence, Hall told me.

"Well, they tried to persuade me, they tried to show me what diamonds actually look like. I got a real education on diamonds. I didn't know much about them." De Beers gave Hall a tour of their famous sorting chambers, and he saw "rooms and rooms of raw diamonds laying the length of the rooms. I mean like pebbles. It was the darnedest thing." As the congressman understood it, De Beers was trying to tell him that there were simply too many diamonds to keep meticulous track of them all, and that labeling each one would be nearly impossible.

Hall met with the president of Sierra Leone, Ahmad Tejan Kabbah, though their dialogue was basically futile. Here was a leader who had lost control of his country. At this point in our phone conversation, the congressman's wife, who was listening in, reminded him of another meeting, the one with Foday Sankoh. Hall said that Sankoh "was the most evil man I ever met." They met at the American embassy. When Hall, along with Congressman Frank Wolf (R.-Va.), stood face-to-face with the rebel leader, Hall asked him why he had murdered and mutilated so many, "and of course he denied everything." After the meeting, Hall walked Sankoh out to where his entourage of soldiers waited for him. There were about forty of them. "They were all armed and they were all smoking pot." The whole encounter gave Hall the chills.

⭑

The congressional impasse lasted about a year. What finally broke the deadlock between the Hall and Gregg bills was television, Smillie told me. In 2001, a Christian humanitarian group called World Vision seized upon the success of the political drama series *The West Wing.* World Vision bought fifteen seconds of airtime, and, at the end of the second season, Martin Sheen, who played President Bartlet on the show, asked viewers to support Tony Hall's bill in the name of "conflict-free diamonds." Even though the ad aired only in the Washington area, its effect was viral. It won Tony Hall the endorsement of the World Diamond Council, and two years later, the Clean Diamond Trade Act was signed into law. The bill's passing in 2003 was sped along because, Hall told me, information "came to us and came to the Congress that Bin Laden was using conflict diamonds to fund some of his terrorist activities . . . and when *that* happened legislation . . . passed pretty quick."

According to Smillie, the Clean Diamond Trade Act was a replica of neither Hall's nor Gregg's bills, but its design was closer to the proposal that Hall, who by 2003 had moved on to become an ambassador to the UN Agencies for Food and Agriculture, had set forth. The act legislated America's participation in the Kimberley Process (KP), the world's monitoring system for rough diamonds, so named because of that first gathering in May 2000 in Kimberley, the same city where, centuries ago, men had opened up the earth and found heaps of the gemstone that would mold the fate of the African continent.

The process was put into action in 2003. It mandated that each international batch of rough diamonds be sealed in a tamper-resistant container and attached to a forgery-resistant certificate detailing the origin of the stones, before it can pass through any borders. The KP is made up of the World Diamond Council, NGOs, and more than seventy countries that have adopted legislation to control where their diamonds come from, to whom they are sold, and what happens to them while they're in the country. Member countries are not allowed to deal with those who don't participate in the process.

I learned from Smillie that sales of conflict diamonds—defined by the process as rough diamonds that fund rebel insurrection against legitimate governments—have dropped sharply. In the middle of the nineties, when Angola's UNITA rebels were still strong, blood diamonds represented approximately fifteen percent of international sales. As talk of legislation spread in 2000, that number was closer to three or four percent. Today, after the institution of the KP and the cessation of the wars in Sierra Leone and Angola, it's closer to a tenth of a percent, Smillie estimates. Of course, that is exactly what all of these figures are: estimates. The thing about illicit trade is that it's rather hard to measure.

Critics of the process, including Smillie, point out that it does not effectively punish infringements and that its minimum standards do not incorporate the monitoring of diamond cutting and polishing. In 2008, PAC noted that conflict diamonds had leaked out from the Ivory Coast and were passed on to customers, instead of being halted by the manufacturing part of the pipeline.

KP skeptics do not consist of only NGO members. Sal Lipiner is among the disbelievers, though his conclusion isn't that the process should be stricter but rather that it was a farce to begin with, just there to "put the public to sleep and say to the public, 'Look, guys, you don't have to worry, don't worry about it. This engagement ring you're buying is not blood diamonds and nobody's arm got chopped off because of this diamond. It's a *wonderful* diamond. Go give it to your bride, she'll be happy, and neither one of you have to have a conscience problem.'"

By 2009, people were becoming disillusioned with the KP. In late spring, Smillie resigned from it. Even Rapaport, who had sung its praises when we met in Las Vegas, said the KP was being used as a "fig leaf" to hide a tragedy, this time on the other side of Africa: Zimbabwe.

What happened in the Marange diamond fields of eastern Zimbabwe sparked an international debate about the definition of conflict diamonds. No rebel forces controlled Marange. Unlike Sierra Leone, the deaths were not a result of war. I learned about the Marange story from conversations with Ian Smillie and Tiseke Kasambala, a Human Rights Watch researcher who traveled to the country, and from a report that Human Rights Watch released, called "Diamonds in the Rough." The Zimbabwean government has disputed many of the atrocities its authorities have been accused of.

It started with a finding. In June 2006, in an arid section of the Marange district, about 360 kilometers from Zimbabwe's capital city, Harare, villagers stumbled upon something that glittered. It soon became clear that Marange contained gem-quality stones. Between 2006 and 2009, millions of dollars' worth of diamonds would come out of Marange.

Shortly after the discovery, authorities associated with the Zimbabwean ZANU-PF government announced that anyone who wished could come and dig. The only problem was that the government did not have the legal right to grant this privilege. According to the Zimbabwe Mines and Minerals Act, Kasambala told me, only someone with a prospecting license can look for minerals, and then, only a person with a mining lease is allowed to actually extract diamonds. Not to mention buying and selling, which require separate licenses.

The government's incentives to let the masses into the diamond fields were probably twofold. First, popularity. The ZANU-PF was faltering. Its national invitation into Marange's diamond source was likely an attempt at mollifying a nation that had become fed up with its leadership. Their second motive was profit. Kasambala explained how party members would take advantage of the anarchy to enter the fields and line their own pockets.

The end of 2006 found Marange in a state of utter disarray. In November, the government tried to undo the epidemic of illicit diamond mining it had invited by sending policemen into the fields. But the operation, which led to the arrests of about 22,500 people throughout the country, about 9,000 of them in Marange, served only to exacerbate the pandemonium. In their surveillance of the fields, the police saw a business venture. Human Rights Watch learned from their interviews that miners would bribe police before entering and leaving the fields. Sometimes the diggers paid in cash, sometimes in beer, cigarettes, roasted nuts, or women. By 2007, if a digger wanted a good place in line, he had to be prepared to pay at every checkpoint.

"In the evening," a miner told Human Rights Watch, "we would approach the police and say, 'We want to work the fields tonight.'" The miners would pay the officers, and when darkness fell, they would go in and start digging. At times they dug until they heard the sounds of gunfire in the nighttime air—their signal to abandon the mines. They carried the ore on their backs. Then they paid the policemen once more and disappeared.

In addition to bribery, the police made their money by intimidating miners into becoming members of "smuggling syndicates," as Kasambala called them. Two to five policemen would band together and recruit a unit of miners, sometimes as many as thirty people. They would grant them access to the diamond fields and then make them surrender part of the profits gleaned from selling the diamonds. The buyers for these stones, Kasambala told me, were from all over the world—Israel, Lebanon, the Congo, Nigeria, South Africa, India, Belgium, and, of course, Zimbabwe.

One man Human Rights Watch interviewed remembered a night in August 2008: "We had decided to go into the diamond fields without paying the police because we had run out of cash. We were digging in darkness when the police fired a searchlight into the sky, and the whole field was as bright as day." The police shot at them. Four miners with whom this man was working got stuck in a tunnel that caved in. The police killed three people and forced the other miners to bury them in one of the digging trenches the next morning. "When I asked to dig out my four colleagues, a police officer told me, 'Consider them already buried.'" Some of the police who were let loose on Marange reportedly raped women who resided on the diamond fields and arrested, beat, tortured, and killed miners.

According to "Diamonds in the Rough," on October 27, 2008, a month after ZANU-PF consented to become part of a power-sharing government, the country's army was unleashed onto the Marange fields, and the mines turned into a bloodbath. It was called Operation No Return, and it began with the flap and buzz of helicopter rotors and truck engines. Tear gas and gunshots rained onto the fields and the nearby towns. Eight hundred soldiers entered the district. The report states that in three weeks, the Zimbabwe Defence Forces killed more than two hundred people. Some of the survivors were forced to dig mass graves. The army introduced full-scale forced labor into the diamond fields, and the miners mined for fear of torture, sometimes at gunpoint. They turned over their findings to the soldiers.

It's like slavery, said Kasambala. As of June 2009, children were still made to lug ore in the sun, sometimes working eleven hours straight without food or water. The Marange villagers lived in fear. And the buyers kept coming for the diamonds.

Because it was the government and not a rebel movement inflicting the violence in Zimbabwe, Marange stones did not fit into the KP definition of blood diamonds. But not everyone thinks it should be like this—Kasambala and Smillie believe that Marange stones are conflict diamonds and that the world should call them that.

In the beginning of July 2009, the KP sent a team to Zimbabwe to investigate. A few weeks later, a draft report was leaked to the media. It revealed that KP investigators believed Zimbabwe should be suspended from the process. Suddenly, there was an about-face. I learned from Smillie that, after the trip, the investigators who had proposed Zimbabwe's suspension caved in to pressure from the South African and Namibian governments, which opposed the action.

Smillie was flustered. The KP is a way of protecting the legitimate diamond industry's reputation. "I mean, this is their livelihood. Frankly, they should be in there screaming and yelling and kicking ass."

From November 2 to November 5, the KP met in Namibia. It chose to grant Zimbabwe six months to reform Marange, despite Israel, Canada, and several human rights groups' support for the country's suspension. Zimbabwe promised to start removing its army from the diamond fields and consented to strict supervision by an independent party.

Not everyone bought it. A few days later, the Rapaport Group issued a trading ban on Marange diamonds. Any member of RapNet caught posting them would be thrown off. Next, Rapaport resigned from the World Diamond Council. He felt the industry's leadership was failing its members by not informing them of the tragedy that was affecting their trade. The average dealer knew little if anything about the situation in Marange at the time. "I mean, why should they know? They've been told the contrary. They've been told and conditioned to believe that if it has a Kimberley Process certificate, it's okay." Some dealers took a proactive approach to avoiding ethically questionable diamonds beyond the demands of the KP by refusing to buy green diamonds (many of the gem-quality diamonds that emerge from Marange are greenish in color), but many were left confused.

I was among them. No matter how much I read about blood diamonds, no matter how many people I talked to, when friends asked me where to buy bargain engagement rings and diamond gifts, as they often did, I didn't know what to tell them. KP certificates say only what country a diamond came from, not which mine. Would I be sending them toward a Marange stone?

By the summer of 2010, the Marange story had picked up a lot of press. The *New York Times* reported on Rapaport's newest trade warning to its members, stating that anyone who deliberately sold Marange diamonds on RapNet would be kicked out. This time, however, they added that the names of these people would be released. This warning came in the aftermath of a decision by the KP in July to certify *some* of Marange's diamonds, those mined in two of the field's concessions during a certain time frame. The stones, if channeled effectively, could provide millions of dollars to Zimbabwe's ailing economy. But the Rapaport Group broadcast to its members that a Kimberley Certificate did not assure diamonds "free of associations with human rights violations," and the diamond world remained divided.

Although diamonds can be used to perpetuate unthinkable sufferings, they also have the potential to lift entire countries from a cycle of debilitating poverty into a sustainable economic situation.

Once the war in Sierra Leone ended in 2002, diamonds from that country were no longer considered conflict diamonds. But Sierra Leone has repeatedly come in last or second to last place in the UN's Human Development Index, which ranks countries according to social and economic development. In 2008, almost a third of its children did not reach the age of five. Rapaport and others plan to bring about a transformation through the stones themselves, and he thinks the answer lies in Fair Trade diamonds, diamonds whose sellers ensure just payment of workers and dedicate a portion of revenues to developing the communities from whose land and by whose diggers the gems were extracted.

Fair Trade jewelry has a real market, which, in the world of buying and selling, makes it serious.

Diamonds, as a commodity, derive their value from sentiment alone. When De Beers marketed the diamond in the 1930s, they didn't market it as a useful object, but as a symbol of love, wealth, and success. In terms of retail, these sentiments are a diamond's only real assets; if they are tainted by the stone's association with war and destruction, the value of the stone literally decreases. Alternatively, if a customer is willing to pay more for a diamond that is not only clean but has contributed to a community in Africa, the diamond gains value, Rapaport told me excitedly, the same way it gains value after being polished or set in a pair of earrings.

Rapaport described an ideal scenario to me. A jeweler starts carrying Fair Trade merchandise in his shop. This jeweler isn't an activist. He's just a businessman, trying to make a living. But soon, "all of the socially conscious, nice women on Rodeo Drive" flood his store. They can afford to pay a bit more to feel good about their jewelry. Fair Trade becomes the new tennis bracelet. The jeweler across the street witnesses his customers streaming into the other guy's shop. He is inspired, in a strictly economic way, to start selling Fair Trade jewelry. The simple laws of supply and demand help spread the product.

The Rapaport Group breaks their Fair Trade criteria into four major parts: fair wages, community benefit (meaning, the communities gain an additional five percent above the fair price), do no harm (whether environmental or human rights), and monitoring and branding.

As of yet, Rapaport does not sell Fair Trade jewelry because there is no recognized third party that certifies it, which is, of course, the big missing link. While there are companies that sell Fair Trade jewelry, like CRED Jewellery and Brilliant Earth, the industry has yet to come up with a Fair Trade diamond.

Why not? I asked Greg Valerio, the founding creator of CRED Jewellery in Britain.

"Because it's so bloody difficult, nobody's prepared to do it." Greg laughed. "It's as simple as that. The diamond trade is going to be the single most difficult thing to certify from a Fair Trade angle." Physical traceability of diamonds, a necessity for Fair Trade monitoring, is effectively unattainable because "organization on the ground amongst alluvial diamond diggers is almost total anarchy."

At its core, Fair Trade is designed to source jewelry that comes from digger cooperatives, communities of small-scale artisanal miners. The industry as a whole is still largely dominated by governments or big companies such as De Beers, though De Beers, Greg admits, came closest to a Fair Trade diamond with their Mwadui Diamond Project, which was discontinued when they sold the mine that would have produced them.

There are other plans for altruistic diamonds, such as Development Diamonds, which seek to return fair profits to miners and provide a safe working environment that violates neither human rights nor environmental standards. The key difference between Development and Fair Trade diamonds is that the former do not require digger cooperatives, a category that excludes most artisanal diggers. The Diamond Development Initiative, which Smillie chairs, sees its work as part conflict prevention; after all, the majority of those who were exploited by the rebel forces to fund their wars were artisanal miners.

Greg Valerio is hopeful that the launch of Fair Trade gold in 2010 will provide a stepping-stone for Fair Trade diamonds. Fair Trade gold introduced a novelty to Fair Trade jewelry—the advent of high end. When I was investigating the movement, Fair Trade jewelry still inhabited the mid-end market in terms of cost. CRED Jewellery all goes for under $20,000. Other companies have similar prices. But it is possible that the more expensive Fair Trade jewelry is, the more it will sell. After all, in addition to their advertising campaign, it was De Beers' manipulation of diamonds and their prices decades ago that made the gems so widely desirable. In the circular world of diamond value, price itself determines how much people are willing to pay for a stone. If Fair Trade diamonds are seen as a genre of high-end diamonds rather than a high-end genre of Fair Trade, they might just catch the attention of people who are willing to shell out thousands, even millions, on a rock.

Of course, this exact phenomenon—the fact that diamonds so captivate people, they'll pay almost anything for one—allows for all the carnage. It is no wonder, then, that there was a time when humans believed something as beautiful as a diamond must be linked to the heavens.

Chapter 13

The Shows

Every few months the diamond men and women take inventory, pack up their goods, and call the armed movers. They ship their merchandise and fly themselves across the country. In January it's Miami, in February Tucson, and at the end of May, sometimes bleeding into June, they deal diamonds and jewelry in the land of bachelor parties, poker tournaments, and five-minute nuptials—Las Vegas. A dealer could spend his entire year buying and selling around the globe at jewelry shows in London, Hong Kong, Basel, New York, Baltimore, Washington, New Delhi, Bangkok, Toronto, Vancouver, and Bermuda. On popular flights to show cities, religious men gather together in the front or back of the aircraft and pray. They keep their voices softer than they would in a synagogue, issuing a low hum of Hebrew words.

In 2008, I joined the dealers in Tucson, a city that turns into one large jewelry bazaar for two and a half weeks each year. Men and women deal out of ground-floor motel rooms and hotel hallways. My father had told me he didn't get any sleep the year before because the dealers in the next room were up all night selling. He voiced this as a complaint, but I was delighted, envisioning a city full of dealers, a larger version of Forty-seventh Street. So I booked a room in a small, rather seedy motel, because I'd heard dealers would be staying there. I got a first-floor room with a window that didn't close properly. I wondered how safe I was, given the fact that the semibroken window faced the road, but I was determined to live among the dealers. As it turned out, most of them had chosen the Marriott or other motels on the freeway. I listened and listened but didn't hear any bargaining through the walls of my room.

I had arrived late in the evening on a plane with what I recognized to be a lot of dealers, mostly because they were Hasidic. At the airport, I spotted Michael Goldstein. He wore a hiking hat, fitting for the desert. When I called his name, he greeted me with a kiss on each cheek, then went off to the Marriott—my parents had booked there, too—while I continued on to my motel.

The next day, I took a cab to Tucson's convention center, where the American Gem Trade Association show was held, and picked up my badge. The antiques were located downstairs in a back room, which is usually a skating rink. Empty audience seating stared down at the arena, and I thought I felt something cold beneath me. When I walked into the room, I was greeted by a dozen twitching apparatuses with wheels. A centrifugal magnetic finisher was cooking up a tornado. Gem shavings lay around in machines and ultrasonic cleaners bubbled. An enlarged television screen showed a man turning a heart-shaped diamond ring between his fingers.

The fact that stone magnification and gem-cleaning equipment were the first things one saw upon entering the room caused my father to feel mistreated. No one even knows we are here, he complained.

It was his second year in Tucson. He hadn't wanted to come originally, but Hartley Brown, the Scottish jeweler, had convinced him. I remember sitting around the Shabbat dinner table in my parents' apartment with the Browns. Hartley and his family had just moved to New York from London, and Hartley was telling my father about the shows he'd been to. In Palm Beach, they had a lounge. "Someone tried to pick up Nicole." His wife. "A gigolo." He laughed.

"Ooh la la," said my father.

Hartley wanted my father to come to Tucson with him. "But don't bring your boys," he said—the Forty-seventh Streeters.

Several times that night, he told my father about all the great customers he'd met there. "You don't want to make any money?"

"I'll buy the Lotto."

But eventually my father relented. A year later, he was in Tucson.

The antiques room was filled with aisle after aisle of booths. On the right side of the hall, I found my father's—a simple affair with black type printed on a white sign presenting the company name, a safe, and before it, a cluster of display cases holding the goods. The first two rows inside the cases were filled with jewelry, and the third was populated with boxes of loose stones. In a corner case, my father kept his prize items: an old Tiffany choker and a crescent-shaped emerald he had on consignment from a man in Germany, which had supposedly belonged to the German kaiser. During World War I, the story goes, a servant of the kaiser's held the emerald in his mouth as he escaped. After the war, he entered a German shop and asked the store's owner, "Are you interested in the kaiser's jewels?" He was. Now it lay against yellow tissue paper—for color contrast—in my father's showcase. But the most exquisite piece was a Belle Époque necklace with thick, tear-shaped emeralds dangling from it like gumdrops. Diamonds formed little loops between the green stones and sat in clusters above them. It was the type of necklace I could imagine a Victorian noblewoman or rich heiress wearing, the type of necklace that stops dealers in their tracks as they march through the aisles of a trade show.

My father had also packed some of his risk pieces, like the warrior necklace, that tacky chain with ivory Hun-looking faces. You never know what might strike a buyer's fancy. Shows are where you meet new faces, new people with different tastes and different markets to fill. Dealers come from everywhere to exchange merchandise, sending the jewels out into the wider world.

My father shared his booth with Lester, the same man with whom he had run from thieves years ago. They've been friends for more than two decades now, and like an old married couple, they are so certain of their mutual devotion that courteousness is no longer an obligation. But before they part for long stretches of time, they always embrace, and sometimes they say "I love you."

When I got to our booth, Lester kissed me once on each cheek. Hartley and Nicole had the cubicle across the aisle, but over the next few days they could often be found at our booth, and we at theirs. When business was slow or no one was looking, my father and his friends chatted or told jokes. They christened Hartley's wife "chéri," because she is French.

My mother was late that day, but I learned from my father that she was in love with one of the items on display—

a white gold ring with diamonds in the shape of bubbles crowding around the center. When she arrived at the showroom, my father updated her on the ring's status. There had been interest. Romantically, he kept raising its price. At a certain point, though, he said, he'd have to settle. He is still a businessman, after all.

When I decided to go see one of the other shows in town, my mother insisted on coming, so together we crossed the large road outside the convention center in the dry desert city and entered a large tent full of merchandise. Not just jewelry but globes, vases, small Buddhas, crystal, jade turtles, a pair of Chinese lovers loving from behind, and a thousand other items.

Here in the tent, my mother could forget that she was at work and pretend to be a customer. On Forty-seventh Street she is a secretary. She works at the desk in the front of my father's office, sorting his papers, systematizing his GIA certificates, entertaining clients, and shielding him from small talk he doesn't want to make. My mother, a Ph.D. in art history, has a creative flair, which she occasionally puts to use by advising my father on which pieces look special to her when he's conflicted about a purchase. Most of the time, her job is limited to running his errands, keeping records of his sales, answering the phone with "Oltuski Brothers" before he picks up the line on his end, and, of course, preparing his diamonds for shipment to the shows. Packing diamonds, she says, "certainly beats packing potatoes in a sack," but when you handle the same stones, necklaces, and rings every day, organizing them, cleaning them, and stringing tickets through them, they lose some of their magic.

Left to her own devices, my mother cannot resist beautiful things, sometimes even taking in other people's discarded furniture from the street corners of New York if she sees potential in them. When we first moved into our apartment, she befriended an Italian painter and provided him with board in exchange for his painting the ceiling of the master bedroom. She covered our floors in marble. After a series of noise complaints from our downstairs neighbors, she was forced to buy rugs. According to building laws, we were supposed to have eighty percent carpeted, but my mother couldn't bear to hide the black and white marble squares she had worked so hard to get installed, so the yellow rugs stood upright in our closet for months. But every once in a while, when we expected maintenance to visit, my family—terrified of being expelled from our residence—rolled out the yellow carpets, just in case. Though their appearance in our home was sporadic, the carpets were beautiful. My mother can't stand ugly things.

In the office, there is little time for her personal tastes. When she's running errands on Forty-seventh Street, she usually can't look at the merchandise the way privates do. But in Tucson, surrounded by sparkling jewels, my mother was a private, and soon she fell in love.

We were standing at a booth whose sign indicated the company was from New Delhi. My mother had her eye on a necklace and a pair of earrings that were quite expensive.

"Ma'am," said the Indian vendor, "if you were to get these made in the U.S., it would cost . . ." His arguments were logical, and if my mother was not already swayed, they convinced her. "Ma'am, both of us know it is worth more than what I quoted you, but I have my reasons."

We talked with the vendors a little bit and learned that they split their time between America and India—six months in one, six months in the other.

"Like Opa did," I said to my mother.

"And also like Papa," said my mother.

"You know, the problem is my husband is in estate jewelry," she told the vendor, coquettishly. What she meant is that he would take his loupe to whatever she bought and investigate it for flaws, that he would see it with the eye of a jeweler, not the eye of a husband.

"What's the problem?" asked the vendor.

When they had been talking for a while, it was clear that my mother had been won over. But she wanted to know "if my husband says, 'Oh my God, how could you?'" could she bring it back? The vendor generously agreed. Then he said that my father should come look at their booth, as well. Perhaps he'd find something he wanted to add to his collection.

My mother explained that we only do antiques, but she invited the man back to our booth.

"You don't do reproduction?" We didn't.

I could already see he wasn't going to buy from us.

"All of us hope your husband will love it," said one of the vendors, "so you'll come back and buy a couple of more pieces."

As I strayed farther from the convention center, the boundaries of formal jewelry selling began to break down. Gems casually took up residence in every crack of the city. Nowhere was this more apparent than on the freeway, which was dotted with motels that had been turned into a makeshift mall of jewelry. Eager to see this market, I headed toward the highway one day, leaving the AGTA's air-conditioning for the scorching sun. I felt like a hitchhiker. From far away, there seemed to be block parties going on. As I got closer, they looked more like flea markets. Only when I went into the motel rooms themselves did the concept of actual jewelry seem to come into the picture. Unlike the AGTA and the tent, where all attendees must preregister, anyone was welcome here.

Enchanted, I walked along the freeway. Outside the ground-floor rooms of the Days Inn, the Howard Johnson, the River Park Inn, and others, lower-end vendors displayed their merchandise on long tables. Tents were set up in motel courtyards, and the smell of incense wafted through the air. Gems the size of babies stood on exhibition, as did singing bowls made of crystal. Canvas bags with large boulderlike minerals sat around on the ground. Here on the freeway, it was easy to remember that the gems were all just rocks.

Inside the rooms, men and women had moved the standard furniture out of the way to make space for jewelry. In one room, two wooden headboards still floated on the walls where beds had stood hours earlier. A comforter and velvet blanket were piled atop a chair. Spanish television ran. The bathroom door had been closed so customers wouldn't see a toilet while browsing for crystals.

Another room I visited housed a Moldavite business. The merchant stood behind a counter, which made the room look a bit more like a store in a mall, and explained to me that Moldavite is a type of meteorite that landed on earth approximately 14.8 million years ago. Soon after I entered, another man walked into the room holding a knife. I got nervous. Security on the freeway was virtually nonexistent. But no one else seemed to mind. The man didn't rob the place. Instead, he just asked the person in charge, "Hey, do you by any chance have the time, sir?" As he waited, he scratched himself with the point of the knife. Then, when he got his answer, he casually walked out. If anyone had flashed a knife at the convention center, security would have pounced on him.

At the River Park Inn, I stepped into a room filled with crystal skulls. Some of the skulls had names. One, whose cranium had been sculpted into sharp points, was labeled "Rock Star." On the floor in the middle of the space sat a gargantuan skull, which the vendor claimed was the largest crystal skull in the world. He asked me if I'd like to take a picture with the skull's carver. For the shot, the carver, a Brazilian man, draped his arms over the top of the statue, dangling his hands onto its forehead, like a father in a family portrait. Afterward, he took me outside to the room's terrace to show me his other work—an entire collection of crystal phalluses. In addition to being a crystal carver, he was also a Lacanian psychoanalyst. I'd never thought of Lacan as someone who could inspire crystal carvings. And I never thought I'd encounter him in my father's business.

A few of the Brazilian man's sculptures depicted thick-thighed, long-haired women straddling giant crystal penises. Others were crystal hands gripping crystal penises. Some of them had been chiseled from clear crystal. In one, the internal grain lines of the gem were so pronounced that, if you looked closely, it seemed as if the statue had been glued together after shattering into pieces. Before I left, the carver gave me his business card, which was printed on photo paper and featured a crystal man in a Thinker's pose. He still e-mails me pictures of his work.

These were the kinds of things my father and his colleagues wouldn't be caught dead selling. But the freeway also included items—black obsidian, gem necklaces, and heavy hunks of semiprecious gems—that wouldn't have looked out of place in the display cases of the Tucson Convention Center where my father and Lester dealt.

I arrived from the freeway during cleanup, and so did an important dealer from L.A. As he browsed our showcase, my mother told me quietly in German that he sells to a lot of big people.

Today's diamond dealers cover all ranks and reputations, but there are still some names that are internationally recognized, such as Lev Leviev, whose rough diamond empire is so extensive it has posed competition to De Beers, and Laurence Graff. They are today's superstar dealers.

The man from L.A. looked at my father's jewelry and asked if he had any nice Asschers. My father must have been nervous. First he said no. Then he remembered that he did, of course he did, he had a beautiful one, and he took it out. The dealer looked at it. He asked how much. Soon, he asked the price again, then answered for himself. My mother and I looked on expectantly. He seemed interested. Suddenly, though, his partner wandered off. The dealer said, "We're walking," and left. In retrospect, I wonder if this was some sort of preplanned signal, like arranging a phone call during a first date, just in case. My father told us that he knew the dealer wasn't going to buy. He continued packing away his diamond boxes. I felt the same pang I had experienced years ago in Munich, watching his merchandise be turned down for the first time.

The next day, a small group of women came to browse

our showcase. One of them asked if my father's merchandise was original. My father pointed to a piece that contained reproduction, but he explained that the majority were original Edwardians and Victorians. The woman said how nice they would be to own.

"You could own it," said my father. They laughed. "You can own 'em all." My father went hunting in his booth for a flashlight to point out a detail in one of his items, but when he found it and turned around, the women were already gone.

"Too many diamonds," said my mother by way of explanation.

"You can take them off." A hint of anger rising in his voice. "Diamonds you can take off." All in all, the show had not been going well. My father had made my mother return her necklace and earrings. He had found shortcomings in them, said they were not crafted well, so my mother dutifully went back to the jewelry tent and gave back the merchandise.

"It's like a museum to them," my father said now. "I should charge an entry fee."

I asked him what his favorite jewel was. He pointed to the necklace with the dangling emeralds. "Except that I've had it long enough, and I'm getting sick of looking at it."

"You know what the problem is?" my mother said at one point. "You need to have *schmonzes* that people can buy as gifts." Cheaper stuff.

But my father doesn't want cheap stuff. He is an idealist. Only, after a while without sales, you get bitter. I recognized it in myself. I was tired of the way people bent down to get a close look at our prize goods, with their feet together, hands behind their backs and heads forward, like birds feeding. You can bet almost anything on the fact that they'll ogle and handle and compliment your goods—goods you polished and remade, photographed and insured—they'll do anything except buy them.

"Is too long, too boring," said Hartley's wife, Nicole, from the across the aisle.

"Sell something," my father said to me on the last day of the show before he stepped away for lunch. My mother had left Tucson a day early to fly back to New York. At a booth nearby, a Hasidic family of dealers had also left their son in charge. But unlike me, he would probably take over the business one day. If someone came along and showed interest in a piece, I would be polite, then call my father's cell phone to summon him.

After my father returned, a woman and her mother came by our booth. I could see from her name tag that the daughter worked at a business called Razzle Dazzle.

"Look," she said to her mother. She pointed out that the diamonds in some of my father's jewelry were quivering slightly in their settings. *"En tremblant,"* she said. She turned to my father to corroborate.

"En tremblant," he repeated. He told me to shake the display case, and I gave it a few pushes from the seller's side of the table so that the *en tremblants* fluttered, sparkling wildly.

The woman told my father she'd been successful in buying this time around. "People come here from all over the country to retire, then they pass on" was roughly how she put it. When they go, their gold and silver and diamonds drift into the market and end up in the hands of dealers, at shows like these.

My father asked about one of the rings on her hand. "Is that for sale?"

It wasn't.

He asked again.

"I sorta paid a lot for it," she said, but she was already sliding it off her finger, uneasily, it seemed to me. I've always thought there was something slightly depressing about selling jewelry right off your body. My mother often used to wear my father's jewelry out for show. When they were at weddings or events, she was his display case, and their friends, his latent customers. I would watch her take her rings off for them to try on, and I was embarrassed for her, embarrassed that the beautiful items she wore seemed to never truly belong to her, that my mother walked around like a human shop, from which our wealthier friends could pick off anything that caught their fancy.

In the beginning, she hated it. Eventually, she changed her mind, because it was nice to wear items she didn't have the money to buy, and also because she saw the women at the Oscars doing it on TV. They, too, would have to give back their beautiful necklaces and earrings at the end of the night, but this fact didn't diminish their beauty, their grace. Sometimes, if a piece is especially lovely, she experiences a little ache upon returning it to my father. But she has resigned herself to the fact that for her, diamonds are not forever.

My father checked the Razzle Dazzle lady's ring with his loupe, examining her jewel for flaws. I said it was pretty.

"I love these," said my father, referring to the style of the ring, but he did not buy it.

As the women walked away, the mother added, "She's got an eighteen carat," which made her sound a bit desperate.

Razzle Dazzle gave her a look. "That's nothing, probably," she said.

"It's interesting," said the mother, to us.

Razzle Dazzle told us she was planning on selling it for scrap, and I felt a tinge of pity, because her mother was trying so hard to advertise her merchandise. Because she had to slide her own ring off of her finger for my father to scrutinize and then not buy. But when you are in jewelry, everything is for sale.

"You're blocking my display," my father said to Hartley, who was standing in front of our showcase.

"I disagree," joked Hartley. "It adds a bit of mystery."

But my father had spent a lot of time strategically arranging his merchandise into what he considered ideal formation. And the antique jewelers were already at a possible disadvantage because of their back room with all the machines. Windexed showcases, booth geography—jewelers and diamond dealers agonize over these trifles.

The trade as a whole was starting to show signs of suffering—the first hints of the recession hitting our market. "It's not a business to come into," Hartley said to me. "It's all over, baby blue."

If he wasn't going to sell much, though, my father wanted to buy. On my last day in Tucson, I went with him and Lester as they visited the tent across the street where my mother bought her set. My father was trying to fill a collegiate-looking ring setting with a gem. First he tried a rubellite, a red-pink semiprecious stone, the color of a hard candy. But the gem was so small, it fell through the ring's center hole.

Toward the back of the tent, we found a friend of ours, an Israeli gem dealer and his sister. The man brought out a blue zircon for us, a triangularly shaped gemstone that was so fiercely turquoise it made Lester and my father coo. My father held his hand out, palm down, and let the stone rest between his middle and ring fingers as though he were taking it out for a fly.

"I don't like semiprecious, but this I like," Lester said.

One of my favorite things was watching my father and his friends fawn over other people's jewelry. Usually, dealers talk straight about gems: carat, color, etc. It's almost sterile, the way a doctor regards a blood sample or an infection. What matters are facts, measurements. But every once in a while, dealers fall in love, too.

At the booth, my father and Lester got to teaching the dealer and his sister Yiddish.

The dealer showed off his limited vocabulary. *"Mach mir a toyve,"* he said. *Do me a favor.* A sarcastic favorite in the language.

"Tu *mir a toyve,*" said Lester, correcting him.

"Ti *mir a toyve,*" said my father, correcting Lester.

I came to Lester's defense, saying there were different dialects. *Tu, Ti,* they were all variations on the same verb. But my father couldn't resist. He called Lester's Yiddish a peasant's Yiddish.

"Sholem Aleichem was a peasant?" Lester asked, flustered, invoking one of the most famous Yiddish writers. They were like two boys arguing over whose dad made more money.

I asked Lester if he was related to Sholem Aleichem.

He just repeated, "Sholem Aleichem was a peasant?"

By the time my flight neared, the two of them had forgotten about Yiddish and were dutifully escorting me to my shuttle. While we waited, Lester looked at his phone and declared, "Market's up."

"Market's up, market's down, market's sideways," mused my father, as though these numbers didn't determine every single deal he made or even thought about making. He had given Tucson one last chance, and it hadn't come through for him. The next show needed to be better.

Chapter 14

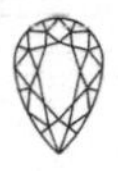

Vegas

Three months later we were all in Vegas, where the burning bulbs of hotel facades flickered like precious gems, where the outdoors looked like indoors and the indoors looked like outdoors. The Flamingo's feathers were Burma rubies, the Eiffel Tower a fancy yellow diamond. The structures that filled the town resembled playthings, as though a child of inordinate proportions had spilled out his toy chest across the city's landscape—a yellow electric guitar, a sparkling pyramid, a white palace, and a glowing blue pirate ship. Hotels lined the strip like vivid embassies of leisure. Splices of Broadway shows and retail advertisements called out from LCD screens; slot machines chimed; and a veil of smoke encased the town. This is the city that often wakes up hungover. But beneath the promise of amusement lies a real desire: everyone is hoping to make it big. In the foggy glow of casinos, where it is always nighttime, people hunch over card tables, trying to turn twenties into hundreds.

The jewelers were gambling, too. They gambled on which pieces to take on the trip and which to leave behind, which diamonds to buy in advance of the show, how much they should ask, and what they'd hold out for. But the big question in 2008 was: What was anything worth? Oil was up and gold was down. And on the Friday before the show, Martin Rapaport had managed to get himself in trouble again. He had released a price list in which the cost of diamonds, especially large diamonds, soared. Prices had been rising over the past several months, but the difference between the May 23 list, published five days before the show, and the previous week's list—May 16—was enormous. On the old list, a round diamond weighing five carats with a D color and an Internally Flawless clarity went for $586,000. One week later, that same diamond cost $732,500. Not all stones' prices had increased this dramatically. The price per carat for diamonds of lesser clarity (for example, a five carater with an Imperfect clarity) stayed the same. But large high-quality stones suddenly became exorbitantly expensive.

This shift was part of a wider development in the diamond market. Diamond mines have been so exhausted that they're producing fewer stones over five carats, making owners of large diamonds, especially large diamonds with great clarity, hesitant to sell. In the months leading up to the Vegas show buyers were paying premiums for these larger stones. Rapaport included a note on his new list, explaining that its high figures did not represent sudden deviations in prices but rather mirrored the premiums people had been paying. Still, influenced by the list, some suppliers were charging premiums on top of the newer, higher prices.

The abrupt change in listings left dealers confused and insecure. During the show, some carried little printouts of the old price sheet in their wallets along with their bills and family photos. Others tried to estimate reasonable values by using a combination of both lists. No one was sure how the new prices would affect trading during the show, but most people were not optimistic. Who wants to buy at such high prices when a stone's worth could drop just as quickly?

The diamond men had seen this type of price increase thirty years earlier. It came right before the big price crash of 1980—their Great Depression. They live in constant fear of a repeat, and since very high prices usually precede very low prices, diamond dealers worry not only when numbers are bad but also when they're too good.

According to Martin, he had no choice. He had held off on increasing figures until he was certain these numbers were not mere speculation, that actual customers, and not just diamond dealers, were paying such high prices. Still, many in the industry were upset that he did it so suddenly, and that he did it the week before Vegas.

It was my first time in Las Vegas. I didn't know anything about Sin City, and my gambling experience was limited to a few hands of blackjack at the cheap tables in Atlantic City. After arriving at the airport, I shared a shuttle with a hyper trio of bachelors. I listened to them talk hotel prices. I had already gotten my bargain—a trade show discount at the Rio, home of the antiques show where my father displayed his goods. They got off at Treasure Island. We were headed to two different Vegases.

The Rio is not the most extravagant hotel in Vegas. It does not boast an indoor/outdoor roller coaster like the New York-New York, or a replica of the Eiffel Tower like Paris Las Vegas. There are no men and women dressed in striped gondolier outfits to hail you a cab or offer you a bottle of water as there are at the Venetian, where the larger JCK (Jewelers' Circular Keystone) jewelry show opened a few days later. When I walked into my hotel, I was greeted by a colorful lounge, a sprawling casino, and a long check-in line.

Each day, on my way to the show, I passed the edge of the casino. Sometimes a lady in a tight sparkling outfit did a dance on a table, but I didn't see the jewelers pay her any attention. They had only jewels on their minds. Still, to avoid glimpsing exposed women, many of the Hasidim on the trip booked rooms at a "non-gaming" establishment. The casino was so smoky that I tried to hold my breath from the time I left the elevator until I got to the restaurant-lined corridor that led to the convention room, but I could still taste the cigarettes when I exhaled, as though I had smoked them myself.

Like Tucson, the convention hall at the Rio is filled with row upon row of booths brimming with jewels. Circulating the aisles and stationed at the doors were armed guards. Each year, the show hires a security force to safeguard the hall twenty-four hours a day. On the first and last days of the show, armed moving companies set up stations in the back of the room. Many of the haulers accompanying the goods were retired NYPD cops or people who had worked part-time as prison guards. Some carried the same guns they did on their other jobs, thirteen-shot eight-millimeter Glocks, but by now I was used to being around men with firearms. I'd seen their vans on Forty-seventh Street, low and squat like tanks, so that no one can get to the merchandise by hanging on underneath, and so the vehicles don't topple over with the weight of their wares. It just seemed as though guns were part of the business.

The jewels had been flown and driven into Vegas from

around the world. From the time the companies took the goods into their possession until they dropped them off in a jeweler's hands at his booth, the merchandise was guarded. Dealers were not even allowed to carry their goods from the stage to their cubicles. Instead, they registered with the movers. Then one or two men transported the bags and suitcases to the booths. At night, the jewels were kept in safes at each booth. When the show was over, the men with guns would return to take the jewels back home.

Wednesday was setup day, and the mood inside the convention hall was industrious. Jewelers and their assistants tore open boxes, unpacked bags, and landscaped their cases. They laid their jewels deliberately across the shelves of their displays, aiming to achieve the Platonic ideal of a showcase—one that would lure passersby to their ware. Hundreds of booths in various stages of readiness stretched out to every corner of the large hall. Bags, duffels, and suitcases were strewn around. Velvet jewel stands of different monochromatic colors stood inside half-filled showcases, waiting to be adorned. Safe doors were left open, while jewelers filled them with parcel papers, boxes of goods, and wads of cash.

As opposed to the rush of precious objects that flowed from the bags, the booths themselves were spartan, composed of a metal framework and separated from one another by black curtains. Ours was a small corner unit, flanked on two sides by the display table, so we had to be efficient about space. My father kept his black leather shoulder bag beneath the table. In it were his GIA reports, his wallet, and a bountiful supply of chocolate nougat Slim-Fast bars, his dieting solution. On the far end of the booth stood a large gray safe, almost twice as tall as the one in our office. Everyone at the show had a variation of this safe. Some dealers had two.

Our booth was more crowded than it had been in Tucson. Lester had brought his daughter Laurie, who was apprenticing with a jeweler in New York. She had come to help her father. Throughout the show, I found myself wondering what it would be like to learn my father's trade in the hopes of taking over the business. I watched the way Laurie tapped her painted fingernails on her telephone to type out numbers and texts, sighed at the end of a long day, or admired jewelry she wished were her own. I realized that for her, this was a normal job. Just like the jewelers, she felt tired at the end of the day and couldn't wait to get out. She was one of them. About a year after the show, Laurie would end her apprenticeship in New York and move back to Los Angeles to work alongside Lester.

There was a third booth partner this time, Richard—a tall and lanky and overtan man with elegant silver hair whom I remembered from my bat mitzvah. And at one point, Roy—"Grumpy"—and his associate, who were attending the show as visitors and hadn't rented a booth, set up their diamond microscope on the table. We hosted an intermittent flow of dealers throughout the show, some there to examine stones, others to entertain themselves while business was slow.

My father opened an old green suitcase swathed in layers of silver tape. It was missing a wheel, but that didn't matter; this suitcase hardly ever touched the floor. After he sealed it up in New York, the armed movers put it in a weatherproof bag, bound it, and transported it to Las Vegas. The wrapping didn't come off until my father held it in his hands. The case was filled with beige boxes, which were filled with diamonds and jewelry.

Lester and Laurie dusted their display trays. They'd wrapped tape around their fingers, sticky side up, and rubbed them against the velvet lining of the trays to trap any dirt. Across the aisle, a pair of vendors snipped shiny fabric and taped it to the glass shelves of their case to make it look fancier. My father cut open one of his boxes with his red Swiss army knife.

He had brought a variety of jewels with him, a combination of ostentatious pieces and what he calls "wearable" jewelry. There were brooches planted with gemstones the color of water and grass; ladies' watches with snaking bands of silver, gold, and diamonds; and a necklace with a broomlike parcel of diamonds hanging off of it, as though its wearer might sweep the table with gems at dinnertime. There was a brooch shaped like a flower with golden, slightly wavy petals, as though the blossom had been caught mid-drift on a windy day. Its interior was a constellation of diamonds so fiery that in one of the photographs I took during the show they glowed in an array of colors, as if the diamonds were not really white but actually multihued, and my camera had caught their secret. Once again, the most impressive jewel was the Belle Époque necklace with the gumdrop emeralds. My father hadn't managed to sell it in Tucson—it was expensive. He would have to charge a lot to make a decent profit. Here in Vegas, it sat prominently on the top shelf of his display case near the corner, where he thought it might garner the attention of dealers from both aisles that intersected our booth.

On setup day each year, before the show opens to visitors, the dealers trade among themselves. They stroll through the room, hoping to find a catch they can take back to their own booths and resell. This time, the day was fruitless for many of the jewelers. My father and another man leaned on our showcase, poring over the Rapaport list together. Oil is high, the dollar is worth nothing, my father said. He had read an article that predicted America would fall just as the British Empire had fallen—the same thing had happened to Britain's prices before the big collapse.

"See that guy?" My father pointed to a man with a shaved head, dressed in an untucked button-down shirt. "He buys people out. He just looks at their collections and says, how much do you want for everything? And you say a million. He says I'll give you $850,000. And he usually wins."

In order to scope out the merchandise, Jimmy Alterman makes his rounds starting on setup day, but he's most successful at the closing stages of a show, when the lion's share of top-tier items have already been sold and all that remains in the showcases are leftovers that jewelers are desperate to unload. He is up front that he wants his customers to be "prepared to take a hit"—Jimmy wants only bargains. Sometimes a jeweler will tolerate a deficit in exchange for ready money he can use to purchase new stock. So, for a price, Jimmy relieves jewelers of their castaways. But he doesn't purchase goods one at a time like most others. He doesn't carefully examine pieces with loupes and measuring instruments, coming back once, twice, with a partner, with a friend, with a fresh mind. That's not his style.

Jimmy Buyout, as he's sometimes called, can glance at an entire showcase and determine its worth in a few moments. He takes a serious look at the whole collection, asks some questions, sometimes touches a few pieces, and then makes his offer.

When he visits a store or a dealer, he intuits which objects are not the seller's favorites. Sometimes he'll go straight for the "safe crumbs," those jewels that people have left in their safes because they don't even think they're tempting enough to set out for customers. "It's all just stuff to me," Jimmy explained when I spoke to him, "so I would rather buy what they're more motivated to sell. It's easier."

What gives Jimmy an edge is that he's not in love with jewelry. He doesn't hunt down the most beautiful necklaces. He's not very interested in the metal latticework of an antique ring. "It's nice to buy a nicer pile of stuff, but it's not always more profitable to do that."

After he goes shopping, he reprices the merchandise. (Occasionally, he gives the owner a chance to repurchase his or her goods, if they feel they've made a mistake.) Then he puts it in his display case and resells the batch. He turns the show into his own bargain basement, and people gather to see what Jimmy got his hands on.

Jimmy is a gambler. He plays jewelry as if it's blackjack. (And at night, when the show closes, he goes to the tables and plays real blackjack.) He doesn't use price tags; the values at which he resells aren't fixed. When he buys someone out, he has in mind a target profit. As he gets closer to his goal, he goes easier on prices. What he charges on a resale bracelet depends on how much he was able to make off a necklace from the same showcase. But he's realistic, and he's antsy to free himself of slow-selling merchandise. So if a show is almost over and it's clear that he won't be making a profit on the case, he'll accept that fact and clear out his newly acquired inventory as quickly as he can for the lowest possible loss.

I found it all incredibly romantic. Jimmy seemed to bring a

little bit of Las Vegas into the jewelry business. Most dealers don't have the capital, the mind, or the nerve for these kinds of games. My father admires Jimmy, but he would never play like him. He understands that the luxury of spending like Jimmy is not in his cards, that he doesn't possess the bravado for this kind of buying. He likes to take his time, considering a piece's quality and salability with care. To Jimmy, jewelry is a product. It can be bought in bulk. To my father, each piece must be justified as an investment. It must be held, touched, and scrutinized before a deal is made.

"This looks like a really busy booth," my father joked the next day. "People eating, looking, everything except business."

The show wasn't all dry for him. That day, I would watch him finally rid himself of the Hun necklace. Roy asked how much he got for it and guessed five thousand dollars. In some professions it might be considered impolite to ask another person how much he made on a deal, but diamond dealers do it all the time. But this time Roy's question gave my father regrets. He had sold the chain for only slightly more than one thousand.

Soon, Lester made a sale, too.

"Wish me mazel tov," he said.

"Mazel tov," said my father.

I took a break from the booth to walk around the showroom; everywhere, I saw the precious stone triumvirate: emeralds, rubies, sapphires. All the other gemstones, my father and his friends refer to as "semiprecious." Laurie told me they aren't called semiprecious anymore. She is part of the new, politically correct generation of jewelers who treat all gemstones equally, labeling none of them just "semi." But my father said, "Everyone in the business still does."

I asked whether diamonds were called "precious."

"No," he said. "A diamond is a diamond."

In the hall, I came upon some odd things I wouldn't have thought of as jewelry, like a pink hair ornament in the shape of a crown, a magnifying glass, and a picture frame. Though the antiques show in Vegas is a high-end affair, some booths had a bazaarlike quality to them, almost like the freeway in Tucson. Black boxes filled with jewels in plastic baggies stood in a showcase, each box labeled with a price—750, 800, 850, 900. They reminded me of cashing in tickets for prizes at indoor amusement parks. String after string of pearls in different shapes and hues lay on a table. Some booths looked as elegant as the store windows along Fifth Avenue, while others were more playful. In addition to the jewels themselves, the dealers dressed their showcases in a miscellany of adornments. In one booth, gemstone necklaces hung from the heads of friendly-looking plastic monsters. On the table of another booth, an animal figurine with webbed feet and feathered hair held a sign that said *Do Not Start with Me—You Will Not Win.* Several people—a family, I thought—stood or sat behind the display case. An older man peered into a baby carriage.

Everywhere I went, I took in snippets of business, of dealers itching to part with their inventories. I overheard a religious man saying, "You know what I'll do?" and proceeding to knock money off a setting, discounting the price of his item.

After I got back to our booth, I asked Lester if that ruby was his first sale, and it was. He told me the piece paid his expenses for this trip. A show can be terribly costly for a jeweler. Some carry millions of dollars' worth of merchandise along with them, which, in turn, can mean thousands of dollars in insurance and transportation. The amount a dealer must reap to break even depends on what he or she earns back home. Someone who earns his living peddling goods on Forty-seventh Street may take a big hit if he closes his office for a week, while the owner of a retail store can keep his business running when he's on the other side of the country or the world. Not to mention, what may constitute good revenue for a single dealer living in Arizona may be inadequate for a dealer with ten children in New York. The priciest jewelry shows are those in Basel, Switzerland, and Hong Kong (the latter requires expensive flights and accommodations, and more time), but even the local shows are risky ventures. Still, they are important affairs, offering dealers thousands of inventories they would not otherwise have access to; thousands of potential buyers to entice with goods that have already been around the block on Forty-seventh Street.

From the safe, Lester produced a small folded paper with the stone he was selling inside. "This is it," he said. I marveled at how unimportant-looking it was. If someone found the closed parcel on the subway, they would probably discard it as trash.

Soon after Lester's victory, though, bad news arrived. The man who bought his ruby hadn't realized the price they'd agreed upon was per carat. He thought the stone cost sixty-five hundred dollars total. He came back to figure out what to do. In the name of honor, Lester ripped up the papers, canceling the transaction.

"How do you not know that?" said my father later, while the dealers gossiped about the incident. I didn't mention that

I hadn't known it was dollar per carat either. Another man thought Lester should have made the buyer pay a penalty for backing out of the deal. There's a camaraderie that emerges when all the diamond men are together. Their body language is the body language of brothers and fathers and sons. Roy leaned on my father's display case. My father told him to move. He was joking, but Roy didn't catch on. He moved, as ordered. My father pulled him back by the arm and softly patted his cheek.

The dealers trade secrets, advise each other, share impressions of other people's goods. My father told Roy that a dealer he met carried "junk." He said he wanted to throw up when he saw the merchandise. The dealers gossip about customers. "She's a *bissel meshugie,*" Hartley told my father, "she's a *bissel tzedret,* to put it nicely." She could be as crazy as she wanted to, said my father, if she bought.

Sometimes the dealers get so caught up in their dealings they forget about others' misfortunes. Later in the day, my father complained about the low price he got for his Hun piece, and Lester said, *"Zei zufrieden"—be happy,* in Yiddish—"I'm still cherry." A sales virgin.

The span of success at a show like this is enormous. For some dealers, it is a financial stretch even to attend. Others come armed with dozens of staff ready to scope out the city for interesting buys. That day, a young attractive woman, an employee of Rapaport, visited my father's booth at the antiques show. I studied her carefully, wondering if she would buy. She was polite but didn't linger very long or ask to see anything in particular. Some hours later, I found myself on a bus to a boat cruise Rapaport was hosting at Hoover Dam. I overheard someone ask the girl about the antiques show. She just made a face, indicating it wasn't worthwhile, and I felt a wave of resentment. The JCK show, where Rapaport was based, was far larger, with more than 3,000 booths to our 264. It occupied several grand and colorful floors of the Venetian hotel and featured all kinds of gems and jewelry, both new and old. At the JCK, logoed carpets and bamboo shoots adorned lavish booths. Still, who was she to make a face at the antiques show? But she was part of the Rapaport Group, the winning team. And for the evening, so was I.

Martin's son Ezi had arranged for me to come along on the trip, but I hadn't been able to reach his cell phone, so moments before departure, I was running through the Venetian hotel, panicking. Finally, I found them in the garage, where they milled around waiting for their coach buses to depart. Many wore red "Rapaport" shirts. The Rapaport gang is composed of people of all ages, but in character, they are young and confident, poised for success, rather than stymied by price lists.

When the cruise boat reached Hoover Dam, Martin stood on the top deck, chatting with those below him good-naturedly. Someone joked about "Rapaport Dam," and Martin joined in, saying, "We've heard damn Rapaport, now we have Rapaport Dam."

By the time I joined the antiques dealers for dinner that night, they were already finishing up. On our way out, Roy asked me if I'd spoken to Rapaport. I'd been feeling a bit like a traitor. When I talked about meeting Martin, I could sense the dealers' envy and curiosity about the man whose price list determined their destiny. I had met the man behind the curtain. That night in his hotel room, I had spoken to Martin Rapaport about everything from De Beers to blood diamonds to man-made diamonds to the Rapaport list. I was very far from where I'd started.

Outside, Roy's son Adam bargained with a limousine driver to take us all back to our hotels. We piled into the car, three on each bench, and flew through Vegas's sparkling nighttime. Roy took out a wad of cash in a rubber band. My father snatched it and pretended to put it in his own jacket.

The show might cost a few of the dealers more money than it would bring in, but they didn't believe in moping. They were together, and they were in Vegas, so they might as well take each other out to nice restaurants and drink good wine, even play a hand at the tables, or, in my father's case, the slot machines. I sat with him one night, enduring the smoke, as he played turn after turn for quarters, while a hotel-sponsored pirate ship filled with half-naked women sailed over our heads, suspended from the ceiling. When he was done, my father handed me the receipt for his winnings.

"Here."

On Friday morning, Lester announced to the booth that they had all become uncles. Another jeweler had had a child. Richard asked if it was his wife or girlfriend. "He runs with a wild crowd," he joked about Lester. "You don't know—wives, girlfriends."

We got to talking about why women tend to go into made jewelry, rather than loose diamonds. My father said women can try the merchandise on, see how well it sits on a female body. "You know, 'It doesn't lay right.'" He gently touched his neck, pretending it was the neck of a woman.

"Jewelry's not for wearing. It's for trading," said Richard.

But my father still believes in diamonds as things of irrefutable beauty, the kind of beauty every girl wants. When another colleague who'd stopped by the booth told the men that his wife didn't like jewelry, my father said, "She doesn't like jewelry? What kind of woman is she?"

"She likes money."

"They all like money," said my father.

The jeweler on the other side of the display said he wished his wife would come along on more shows. Most of the men were here on their own.

It was the middle of the show. Day three. That day my father's heart leapt. He found out that a dealer he knew had just sold a diamond similar to one of his own: a possible pair.

Finding two matching stones is a diamond man's holy grail. It means he can make a pair of earrings. No two diamonds are exactly alike, and in truth, most customers aren't able to discern minute inconsistencies between the stones in their earrings, but the diamonds must appear identical to the naked eye, and ideally to the microscope, as well. Even near-duplicates are difficult to find, so dealers search long and hard for stones of comparable qualities. A diamond with a partner is worth more (sometimes by ten percent) than a solitary stone. So when my father heard about a possible soul mate for his diamond, he asked who its new owner was. When he got a name, he hurried over to that booth and said to the buyer, "I know you bought it, I don't know for how much." The man told my father that he had already flipped the stone. My father told the man to get it back; he would make some money. The man generously replied that my father didn't need him. He gave him the buyer's name. So my father found the second buyer. He explained the situation and borrowed the stone. He walked briskly back to his booth with the open parcel paper in his hand, and took his own diamond out. But when he compared the two stones, the crown was all wrong. At first, I couldn't see the problem. He held it up to me, and then I realized that the slope between one stone's table and girdle, like the hill of a volcano, was noticeably steeper than the slope between the other's. The diamonds were two different shapes, not a perfect match. An air of disappointment settled over my father.

A pair of diamonds is something all dealers desire, but antique jewelry is a far more subjective field. What one man would give his entire inventory for, another wouldn't let into his showcase. Later that day, a young Hasidic man my father knew came by our booth and glanced down at our pieces. He asked if people bought them.

"They better," said my father. "Otherwise I'm in trouble."

"Something like this," said the Hasid after a bit. "Who in their mind would buy this?"

My father said that he would. He explained that, with jewelry, one must consider the workmanship of a piece, its history, its uniqueness.

The Hasid wanted to see an expensive antique jewel, so my father showed him the emerald necklace.

"What means expensive?" asked the Hasid in Yiddish.

My father told him its cost. He noted that the piece was signed, a feature that drives up the price of antiques.

The Hasid asked who signed it.

"T. B. Starr. You don't know the name."

"Some drunk," said the Hasid. T. B. Starr was a famous jewelry designer.

Over the course of the conversation, my father and the man spoke about an e-mail Martin Rapaport had recently sent out. That name seemed to ring from the lips of every diamond man in Vegas. The list. The prices. The economy. The industry was changing, and it was terrifying, exasperating. Who knew what was to become of all these jewels?

But no matter how badly his goods need to be sold, my father never deals between sunset on Friday and sundown on Saturday. Before he leaves, he brings any pieces he owns jointly to the co-owners' booths. The rest of his diamonds and jewelry go into the safe. That afternoon, he put a sign on top of his showcase that read *CLOSED IN OBSERVANCE OF SABBATH Please Return Later*. A few other booths had the same notice.

On Shabbat, two Persian dealers converted their hotel room at the Rio into a small synagogue for prayer services and meals. They ordered a massive amount of food from a kosher supermarket section in Las Vegas, and all were welcome to join. A motley group of Hasidic, Israeli, Persian, and other assorted Jewish diamond dealers gathered there. When I arrived, the men were already praying in front of the television, near the beds. A handful of women stood or sat in a separate cluster by the door. They shifted to make room for me. Throughout the service, it was our self-appointed task to close the door every time someone new came in, so that it would be done swiftly and quietly to avoid the infiltration of outside noise into our quiet little prayer room. I was used to standing with the women during prayer. The synagogues I grew up going to in New York and Germany had a separate balcony section for the women, the idea being that men should not be distracted by female allure while trying to concentrate on what was supposed to be a conversation with God. In Frankfurt, my grandfather used to look up from his spot in the pit of men when I came into temple and motion for me to sit next to the rabbi's wife. I didn't know the rabbi's wife personally, but I was going through a phase of religious piety acquired in Jewish day school, so Opa assumed we would get along. Here in the hotel room, there was no rabbi, just a collection of dealers praying on the Sabbath.

Food lay by the sink outside the bathroom and on the marble floor beneath it, in foil trays and bags. The Challah bread sat on a folding rack meant for suitcases, and in the closet, on the shelf where hats are usually kept, lay a Torah scroll covered in a prayer shawl. One of the dealers had borrowed the Torah from the local Young Israel, in exchange for pledging contributions. One year, my father told me, the Young Israel remarked on the modest amount of the donation, so this time care had to be taken to raise proper funds. To do that, dealers would auction off the prayer duties, which are considered an honor—reading the Torah aloud, dressing it, saying the blessings, and carrying the Torah back to its ark (the closet).

That night, at dinner, the dealers spoke about how the Vegas antiques show would fall on Shavuot the following year, which meant observant Jews would have to miss two days of trading plus the usual Saturday for Shabbat. They talked about complaining, about signing a petition. One of the hosts of the evening, a tall, lanky man so shy and unassuming he almost always smiles when he speaks, announced, "We have to be united."

The second host was shorter and sturdier, with deep brown eyes and round scholarly glasses. Back in New York, the two men share a luxurious window booth at the 10 West exchange. But here in their hotel room, they kept things simple; their goal was just to have enough room and food for everyone. To get to his seat at the end of the table, near the window, the tall shy host climbed over the two twin beds so he wouldn't disturb any of his guests.

At the end of the night, many of us climbed the staircase back to our rooms so that we would not violate the Sabbath by pressing the elevator buttons. The group reconvened for morning prayers the next day. One man stood in the front of the room before they took out the Torah, and listed the tasks. The congregation called out its pledges; even here, in synagogue, the diamond men and women made deals. At some point during services, the curtains were opened and we all prayed facing Caesar's Palace and the Bellagio Hotel, Vegas's temples.

"Look how beautiful that is." A man and a woman were ogling my father's case. It was Sunday, the last day of the show. My father and I smiled at each other. In German, he told me that when someone says something like that—*very pretty*—you know he's not going to buy. But often, as the show nears an end and people get desperate, deals move quicker. One dealer told my father, "I sold a real nice brooch right off my jacket." When you sell something so fast, she said, you start to doubt. You start to think you didn't sell it for enough. In fact, she said, she probably didn't, considering the new price list.

Then it was my father's turn. Three Indian men came by our booth, and a large, round yellow Cape stone caught their eye. Its table was about the size of an M&M. The men passed around the stone. My father gave them a color card, an industry trick that's been used for decades to tell how yellow a diamond is against the backdrop of white paper. The cards are folded and propped up in such a way that they create a little valley for the stones to rest in. One of the men brought the colored paper, with the Cape diamond in it, close to his mouth. He breathed onto the gem, then put it back under the diamond light.

Later on in the day, the men came back and named their price, which my father promptly rejected. So they asked for sixty days—they would pay his price if they could have two months to do so.

My father said he wouldn't take sixty days, that when he buys, he pays immediately. "I give a check right away."

The men replied that this is how he manages to pay lower prices. But now my father was naming the price. In other words, they should get to set the terms. With diamonds, time is almost as important a bargaining chip as price, because it gives dealers the opportunity to cash in on some other sales before paying out.

My father stood firm for a bit, but then he said he'd make a quick call later on to check with Roy, whom the men said they knew. This was the beginning of the end of negotiations. After some back-and-forth, my father and the men shook hands. Presuming Roy attested to their honor, they had an agreement.

My father called Roy. They're okay? he asked. They were, said Roy. Richard, my father's second booth partner, also chimed in that the men were okay.

Before they returned to close the deal formally with *Mazal*, I listened to my father and Richard chat about them. The men had been to our booth over and over again, in different combinations. Sometimes all three of them were there, sometimes just two. My father joked that they'd handled the stone so many times, they "prolly rubbed off a point already." Better polish, quipped Richard, as though the men's fingers had burnished the stone. When the Indian men came back (only two of them this time) they wrote a record and said *Mazal*.

Less than an hour before the show ended, another jeweler dropped by. He knelt down before our showcase. He asked my father about a moonstone ring behind the glass, a milky domelike gemstone that flickered shots of blue perched on a gold ring and encircled by diamonds. My father gave him a price, then knocked a thousand dollars off of it. It was probably his last chance for a sale.

"If I keep kneeling, will it keep going down?" the man asked.

Caught up in the heady game of bargaining, I told the man to lie down. He didn't, but my father declared another, lower number anyway.

After browsing some more, the man wanted to see the moonstone again. My father held the ring out on his pinky finger. The dealer slipped it off and said he would buy.

Soon, the armed movers came to my father's booth with bags and slips. The dealers were packing up their goods like carpet salesmen, getting ready to move out. My father unplugged his fluorescent light and wrapped the wire around it, putting it back in its white container. He unlocked his display case and stuffed his little black boxes of stones next to one another in a carton. A loudspeaker announced the end of the show. One man yelled, "Yes! Yes! Let's go home." The room buzzed with the sound of tape being pulled and stretched across boxes. I watched the movers pull dollies across aisles, some empty, others with bagged items on them. From one of the velvet display necks, my father carefully lifted the big emerald necklace. It had sat in his display case all show long. People had looked, but no one bought. He enveloped it in bubble wrap and laid it in a special necklace case, purple satin on the inside, black velvet on the outside, with two little bands to hold the piece in place.

"It's such a relief when the jewelry goes," said a New York dealer who had come to visit us. "Both in New York and here. Because you know it's safe. It's the only time you don't have to worry about it."

"Stay there," my father said when he stepped from his showcase to his safe, just a few feet, to retrieve items. I stood on the outside of the counter and guarded the open display so that no one could reach over the glass box and snatch something.

A man came to unplug our booth's overhead lights, my father and Lester quarreled, a dealer rode around the room on a Schwinn scooter, and soon, it was time for my father to close his green suitcase, lock it with a key, and tape it vertically and horizontally, adding to older layers of silver tape. From the table into his open hand, he swept the little scraps of paper that had stood in the showcase and said things like *boucheron* and *French,* describing the jewelry's make. In his shoulder bag, he packed his calculator and the remaining Slim-Fast bars. He double-checked his tissue paper for stones, then threw it out. Before leaving, he got down on his hands and knees and investigated the floor for lost items.

After the show, a horde of dealers convened at one of the hotel bars. They filled the area, chattering and reveling. I sat down to a short dinner with my father, Lester, and a few others, but eventually our group began to thin as the men went back to their lives in New York, Los Angeles, and Canada.

Vegas hadn't been a complete disappointment. My father sold a few large items, including the big yellow Cape stone. But he wished he'd been able to sell more, so that he didn't have to stare at the same green gumdrops year after year. No dealer likes to carry stale merchandise. "Listen," Efraim Reiss once put it to me, "we are movers. We try to sell and get rid of the old inventory."

Not everyone went home that day. The JCK show was still going on, and I was staying to go to the annual Rapaport State of the Diamond Industry address, where Martin would tell the diamond people where he thought the trade was headed, and also where the dealers would have a chance to tell Martin what they thought about his new list. It would be a face-off, but despite the stakes, the mood in the high-ceilinged, brightly lit room was sociable that Monday morning. Various members of the trade stood in groups and chatted, but I didn't know anyone. None of my father's close friends were in this room. Many of them had already left Vegas. Although they were all in the same business, these people seemed somehow more corporate, more serious than the dealers I knew well. The men, for the most part, wore suits or button-down shirts.

Bagels were set out at one end of the room, and a short stage with a wooden podium that said *Rapaport* graced the other. Behind the podium stood Rapaport, the man. He wore a black suit, a red bow tie, and a black *kippah,* which, by the middle of the talk, had migrated to the side of his head because of how vigorously he moved around when he spoke.

Two large screens on either side of the stage projected PowerPoint slides, airing messages such as "Free, fair, open, competitive markets" and "De Beers is not your daddy." Also to the side of the stage, leaning on easels, were blown-up pictures of Sierra Leoneans that one of Martin's sons took during one of their trips to the country. Most of the seats were filled. Many people had brought pen and paper to take notes.

"Good morning, everybody. Okay. Good morning," Martin said to the crowd of about two hundred, already moving on his stage. It was just a few days after I'd met him for the first time in his Venetian hotel suite to talk about the price list and Fair Trade. He sounded more awake today. He could yell when he got excited. He had a stage, literally, on which to dance. He had an audience. But before he got to diamonds, Martin offered a prayer. "First of all, I pray that, as I say the things I'm gonna say today, I don't offend anyone, that I don't give people bad advice. We're in turbulent times." He acknowledged the American troops abroad. The audience applauded. Martin nodded. "It puts the diamond business into perspective a bit. I wanna say thank you. I wanna say thank you really to our people, to our clients, to the guys who support Rapaport even when they're pissed off at me."

Soon, he began to describe an economic world that was virtually unrecognizable from the one in which dealers had been trading for years. He told the diamond people that they were "a Ping-Pong ball in a sea of change." He told them they needed a gestalt and then he defined the word "gestalt" for them. He gave them a lesson in recent world history. He unfolded before them the story of short-lived American opulence and how America's spending habits had catapulted its wealth across the ocean to places like India and China and spawned a new middle class. "All that money is going someplace," Martin said, "and those people that are getting that money could buy diamonds like you guys buy potato chips."

The relocation of wealth has tremendous consequences for the diamond industry. A shift in demand inevitably means a shift in supply. And so "competition is not the store across the street. Your competition is some guy with a strong currency in India that can buy those diamonds and has better customers for those diamonds than you do, who will buy the diamonds instead of you."

Sometimes, to lighten the mood, Martin peppered his talk with jokes. "I'm waiting for negative interest rates. I go to the Federal Reserve Bank, and I say, 'Guys, I'll take a loan. Gimme five percent a year.'" Laughter.

He speaks with dashes and parentheses and consults his audience on cultural references; he is too busy making grand points to rescue details from the tip of his tongue. Often he refers to himself in the third person by last name, as though he is a coach in the middle of a tense fourth quarter. And from within what may sound like ramblings to those who aren't listening carefully, there emerges wit, mastermind, and sometimes, terrifying revelations.

"By the way, what year is it this year? Who could tell me? What year? Year of the mouse, yeah, that's cute. Good. Good point, but what year? What numerical year is it? 2008, most people will say. Okay, Rapaport is saying this year is now 1978." Someone in the audience groaned; 1978 was two years before diamond prices crashed.

"Someone said, 'Ohhh.' Right. Well, seventy-eight was a good year, seventy-nine was a good year, eighty was a good year for the first half, and then everything hit the fan. So? What do the Romans say? Party today 'cause tomorrow we're toast?"

Bad economies are not bad for everyone. They inspire innovation, like Jewelers Alert. They provide markets for secondhand dealers, since the first thing a floundering population will sell to access cash is jewelry. It was what Razzle Dazzle had said. Martin told the audience about a friend from Dubai who had been buying up jewelry by the case. "He was goin' nuts. 'Man, I went to that estate show. I went over to a showcase and said, I'll give you three million dollars for everything in the showcase. You got four minutes.' The guy sold him. The guy sold him, okay?" I thought automatically of Jimmy Buyout. "There's deals, man. If you know, if you've got the animal instincts of the marketplace, if you're a hustler and a haggler and a"—Martin growled hungrily—"you know, you got that juice in you, okay? So the Great American Slowdown for an estate buyer—great! Bring it *on!*"

Martin knew he would eventually have to acknowledge the twenty-carat elephant in the room: the list. So he explained the reason behind the chaos in prices. He told the diamond men and women about the shortage of large stones, how investors abroad were buying them up in staggering quantities. Lower-clarity goods were not experiencing the same jumps in price because new international power-buyers were finally able to assert their preferences over the market.

Before the end of his talk, Martin aired a clip of his visit to an alluvial diamond mining site and a diggers' assembly in Sierra Leone. The audience applauded, but Martin didn't react. He was busy fiddling with the computer that connected to the projector screens. As he delivered the last bits of his talk, his voice was calmer than it had been. He wasn't yelling at the room. He wasn't jumping around the stage. He finished his talk and thanked everyone. And then he took questions.

The first couple of questions were pretty friendly, but soon things got more heated. People were cross with Martin, not only for the new price list but because of a subsequent e-mail he sent out to subscribers advising them to look for new suppliers if theirs were charging premiums over the newer, higher lists. He had done so to avoid chaotic speculation based on his new list, to remind people that, as buyers and sellers, they still had a choice in prices.

"Who are you to tell us to go find different suppliers?" one man asked.

Martin would later apologize for telling people to change suppliers.

"Let's call a spade a spade," said another man during his time with the mike. "You're powerful. Your list is accepted all over the world . . . and that power comes with an extreme responsibility"—Martin nodded deeply—"and my personal opinion, and most people in the industry, feel that you've failed to live up to your enormous responsibility that affects the livelihoods of most of the people in this industry."

That day, I got a little taste of what those fierce times at the Diamond Dealers Club must have been like in the early days of Martin's list. But whether the diamond community likes it or not, he has made himself indispensable. "I always say the world is round for two reasons," Martin declared that day in Vegas. "One reason the world is round is 'cause what goes around comes around. . . . The other reason the world is round is because you can't push Rapaport off."

Chapter 15

The Diamond Growers

The diamond growers place the diamond seed in a small compartment. They make sure the methane and hydrogen gases reach the box, check its temperature and pressure, and through small windows in the box, they watch the diamond swell like rising bread. After a few days, they take the diamond out, slice off its impurities, and put it back in, hoping for better results.

The Geophysical Laboratory at the Carnegie Institution of Washington, D.C., resides in a beige-brick edifice with reflective windows and a slanted roof, situated on a grassy hill just past the reaches of public transportation. I take several buses and subways to meet Dr. Russell Hemley, the director of the laboratory. It is exciting to think that diamonds are being grown just a few miles from my apartment.

I am met by Dr. Hemley's assistant, who takes me to the scientist's office. Dr. Hemley is a tranquil man with disappearing eyebrows and hazel eyes outlined by semiframeless glasses. He was part of the team that, in 2000, grew the first gas-generated single crystal diamond of gem quality.

Single crystal diamonds are the kinds of diamonds that are sold in jewelry stores. As opposed to brittle polycrystalline diamonds, which usually get destroyed in nature, single crystals are stronger, clearer, and prettier. But Dr. Hemley is not in it for the looks. He's also not in it for the money. He's not looking to sell his expertise or technology, though many have offered to buy it. Dr. Hemley is in it for the pressure. Force two diamonds together in just the right way, and they will re-create the pressure levels inside the earth's core or in the oceans on Jupiter's moon Europa. Using diamonds, Dr. Hemley and the other scientists can subject living organisms to the rigors of foreign geological environments. When the organisms survive, they demonstrate the possibility of life in these places.

I sit at a round table with Dr. Hemley in his office and start to probe. "So would your research lead you to believe that there is life . . ." I trail off so my question doesn't sound too sensationalist. Dr. Hemley is not even a little bit of a sensationalist.

"No, it just says that life as we know it has greater adaptability to survive, be viable, in those kinds of environments, but it doesn't say that there is life on these other planets or—"

"Do you think there might be?" I want to hear him say it, but he won't let me push him.

"Ohh," he almost groans, "well, that's an observational question. I mean, based on known biology, there's nothing preventing it. It's a matter of what we've discovered from our laboratory experiments that it's not precluded." I let off. I knew he wouldn't cave.

✡

A diamond anvil cell—the instrument that carries out many of the pressure experiments at Carnegie—looks a bit like an hourglass. The pointed bottoms of two diamonds converge in the middle. Before experiments, the anvil is unscrewed into halves. With a needle, the scientists apply specimens onto the tips of the diamonds. Then they reassemble the anvil and set it under the microscope for observation.

Diamonds are the perfect facilitators of high-pressure experiments. Because they can handle high intensities of force, they transfer pressure onto whatever materials lie between them, rather than cracking—at least most of the time. Since diamonds are transparent, scientists can observe the materials on which the pressure is being applied.

The eventual goal is for Carnegie to be able to produce as many diamonds as it needs for experimentation at a reasonable expense and pace. For now, the lab still relies on Forty-seventh Street and other diamond centers for its supply, though scientists have used a few of their own diamonds, as well. Diamond growing is still a pricey and unpredictable endeavor. The costs of mining a diamond from the earth vary wildly, depending on the geographical location and conditions of the mine, but for colorless stones above one carat—the most useful kind of diamond for high-pressure experiments—it is more expensive for labs such as Carnegie to produce them than to purchase them from the trade.

At Carnegie, I meet the laboratory's main diamond grower, Dr. Chih-Shiue Yan. It is his job to perfect the lab's diamond-production technique. His goals include stronger, clearer, and whiter stones. Color and clarity are important, because they determine how much light will pass through the diamonds onto the experimental materials and also how easily the subjects can be observed through the microscope without the stone itself interrupting the view.

Dr. Yan came to Carnegie in the nineties after leaving a position in theoretical calculations, and he has been growing diamonds ever since. From the way he proudly introduces his laboratory colleagues to me, the way he conjectures that diamonds may one day replace silicon, the way he peers into the future of diamond growing to a time when the stones they grow will be "big, big, big, big, big," it's clear that he genuinely loves his job, that the idea of pushing the bounds of diamond creation—thicker, taller, faster—excites him. During my visit, he shows me the lab's creations. From a small safe, he retrieves a few boxes, some of them similar to the boxes I've seen in my father's office, and lays the diamonds out on a table.

They come in various colors: yellow, pink, green, colorless. The diamonds created at Carnegie emerge looking like chips of glass or plastic, but the stones that have been cut and polished appear natural. If I hadn't been told, I would not have known the difference. The reason the lab polishes them like commercial diamonds is that the tip of the stone must be very small for high-pressure experimentation. A force exerted upon a smaller surface area will yield more pressure than one dispersed upon a large surface area. The tips of these diamonds are about as thick as a strand of hair. For the scientists, the most useful part of a diamond is the part most wearers don't care about. In jewelry, the bottom tip usually disappears into settings.

I notice that one diamond is cut and polished into an emerald cut, a rectangular shape, which lacks a point. What about that one? I ask. Why is that useful?

Dr. Yan laughs. "We just wanted to make it beautiful." I find this touching. Dr. Yan is an aesthete. Looks don't usually mean very much at the lab. Still, scientists and jewelers care about many of the same properties. Both want their diamonds white, pure, and hard, but for different reasons. For dealers, color and clarity means luster. For scientists, it means visibility under the microscope. And while a regular wearer of diamonds needs her stone to be hard enough only to survive handshakes, showers, and the occasional drop to the floor, a scientist needs his or her diamond to endure cosmic pressures.

For every compliment a dealer could pay a diamond, scientists have two. In an e-mail to me, Dr. Yan sang the stone's praises: "supreme hardness, high thermal conductivity, low thermal expansion, chemical inertness, excellent optical, infrared, and X-ray transparency and semiconductor properties." Each property has a significance. It is important, for example, that diamonds are chemically inert because it means they won't react with the materials they're crushing in the anvil. Dr. Yan has all of these properties in mind when he works on diamond growing. The stones Carnegie has been creating are already stronger, purer, and harder than natural ones. Their weight is impressive, too; they've created a ten-carat diamond.

The clamor in the diamond-growing laboratory is like rocks being put through a blender. On one side stand a polishing wheel and a sink. The area looks like a movie rendering of a mad scientist's lab. A network of metal tubes runs along the ceiling and connects to large gas-filled boxes. Between them is a slightly smaller case, where the most important action occurs. I find it enormously pleasing to the eye, mostly because the plasma that produces diamonds is green, and so a bright green mist blazes inside the creation chamber, as though something otherworldly is trying to manifest. The holes to the chamber's interior are covered with sunglass lenses, so as not to harm the eye of the beholder.

The diamond-growing method Carnegie uses is called CVD or MPCVD, chemical vapor deposition or microwave plasma chemical vapor deposition. It produces larger and purer diamonds than its predecessor and alternative technique HPHT (for high pressure high temperature), and does so more economically, according to Dr. Yan. MPCVD involves two gases: methane and hydrogen. The gases travel from two separate chambers, through stainless-steel tubes near the ceiling, and into a small box that is both a vacuum and a microwave. There, a small diamond seed—a natural diamond—rests. The diamonds Carnegie grows still start out with a mined stone. Inside the chamber, warmed by microwaves to over eleven hundred degrees Celsius and subjected to a relatively low pressure of below one atmosphere, the gases become plasma, and attach themselves to the diamond seed. Methane contains carbon, diamond's building block. The diamond grows.

This whole enterprise can take a few weeks. And it's not a one-shot deal. It's a little bit like baking. They leave the diamond in for a day or two, check on it, laser off what they don't like, then put it back. Impurities in lab-grown diamonds typically consist of carbon that has transformed not into diamond but rather into graphite or soot, due to imperfect temperatures or gas pressures, or even defects in the seed diamond itself. The scientists repeat the process as many times as they need to. When everything is done, they cut off the natural seed diamond with a laser, so that only their CVD diamond remains.

After showing me the creation lab, Dr. Yan takes me to their modern substitute for a cutting factory, where Carnegie keeps its laser. Proudly, he shows me various shapes the lab is capable of cutting a diamond into. Every once in a while, the scientists seem to remember they're working with the love stone. Romance takes over. They cut a diamond into a heart. One of the students at the lab, Joseph Lai, lasered *I Love You Courtney* into a stone for his fiancée on Valentine's Day. The dedication will never be erased, since more diamond was grown on top of the inscription, thus surrounding the text with stone. This is the scientists' own version of forever. Courtney didn't get to keep the diamond, though—lab property.

Toward the end of our meeting, Dr. Yan arranges for me to look at materials in a diamond cell anvil. The microscope I peer into shows a black point with some white surrounding it. I first assume that I am seeing the diamond in addition to whatever substance they're observing. But instead, I am told that the black is chromite, the white, helium. I am looking right through the gemstone at the gases; as Dr. Hemley and Dr. Yan promised, the diamond is so transparent that, under the microscope, it is invisible to the eye.

Dr. Yan never rushes, never reminds me that there are far more important things he could be doing, such as helping to determine the highest pressures under which bacteria can survive. He never reminds me that he is a world-class scientist. It was Dr. Hemley who informed me that Dr. Yan grew that first gem-quality CVD diamond in 2000. And in 2002, he and a group of Carnegie scientists generated a diamond that grew one hundred times faster than diamonds created by traditional MPCVD synthesis. It emerged over three millimeters thick. That's enough gem to yield a polished quarter-carat round brilliant stone, a stone that would do just fine on Forty-seventh Street. To an average consumer, it would look like any other mined diamond.

Dr. Hemley is just as modest about his accomplishments—almost fanatically so. Although it would be pretty safe to say that, in 2000, the lab at Carnegie, together with the University of Alabama at Birmingham, created the first single-crystal CVD diamond, Hemley likes to modify this statement. He'll only go so far as to say that when the lab at Carnegie made their first single-crystal CVD diamond, no one else had announced that they had—to his knowledge.

Perhaps his caution stems from the fact that diamond growing has a fraught history. In 1955, General Electric became the first official creator of man-made diamonds. But they weren't really the first to produce the gemstone; they were just the first to declare their feat. Two years earlier, a company called ASEA in Sweden had already created diamonds. ASEA decided to perfect the process before going public with their achievement. As a result, they missed the boat, and the distinction was bestowed upon GE. But Dr. Hemley believes that neither was the pioneer, that the first man-made diamond was probably born before the 1950s. According to him, there are claims going back to the dawn of the twentieth century.

Mankind has been trying to create diamonds for ages. A *Harper's Magazine* article in September of 1859, entitled "Something About Diamonds," remarks that "alchemists and chemists have devoted more time and study to this point than any other, only excepting the search for gold in the crucible; and still the product is nothing."

As late as the 1300s, alchemists believed that male and female diamonds could breed and bear little diamond babies. The carbon composition of diamonds was unknown until the late eighteenth century, and ever since, scientists have struggled desperately to create them. Not surprisingly, alongside man-made diamond history is a rich procession of diamond impersonators. The same *Harper's* article announced that, although no scientists had triumphed in genuine diamond creation, "the manufacturers of false stones are improving daily, until the excellence of their wares is so great as sometimes to stagger the judgment of the best connoisseurs." Apparently, Paris was the center of this enterprise and gave its name to some of the more compelling replicas, which people dubbed Paris brilliants.

Today, the company Charles & Colvard produces a gem called Moissanite for consumer wear. Technically the mineral exists naturally in very small quantities, though not enough to craft jewelry, so Charles & Colvard engineers it in a lab. The mineral is composed of one silicon atom and one carbon atom, hence its scientific moniker silicon carbide. Charles & Colvard markets Moissanite as a distinct "jewel" instead of a diamond substitute. In fact, according to them, it possesses more fire than a diamond and is almost as hard.

Nevertheless, though Moissanite is not produced with the goal of masquerading as a diamond, it can still fool a dealer at first glance. Upon closer inspection, however, its facets are double reflective; under the loupe, each facet looks like two.

Then there's cubic zirconium, the cubic form of zirconium oxide. It also resembles a diamond to the untrained eye but, unlike Moissanite, does not deceive most jewelers with loupes and scores far lower than diamond in hardness. It's also heavier.

Man-made diamonds are a different story. They are not fakes; they are not alternative minerals. They are the real thing. And without sophisticated technology, not even a dealer would be able to tell the difference. It is probably safe to assume that a small number of man-made diamonds have made their way into the market without anyone's realizing it, but Dr. Yan figures this number is very small, somewhere around 0.1 percent, because there simply aren't many companies out there creating diamonds, and those that do have not started to mass-produce.

What still amazes me is that, barring costly gear, neither shoppers nor most jewelers could tell the difference between natural and man-made diamonds, but this almost invisible difference renders lab diamonds cheaper. I'd be lying if I said it didn't scare me a bit, as well—on behalf of my father and the dealers.

Labs produce approximately one billion carats of man-made diamonds annually. Dr. Hemley, however, doesn't think synthetic diamonds are a threat to the natural industry just yet: "At this point it's not a problem for them, but it scares them." It scares them so much that the Jewelers Vigilance Committee and other trade entities got together in 2006 to petition the FTC with the goal of disallowing sellers of man-made diamonds from using the word "cultured" when advertising their gem.

The FTC provides vendors of lab-grown diamonds with

four words they can choose from, one of which must accompany the word "diamond" when they advertise. They are "synthetic," "laboratory-created," "laboratory-grown," and "[manufacturer-name]-created." But the petitioners wanted to make sure that, even when producers used one of the four allowed words, they couldn't add the term "cultured." In 2008, the FTC denied the petition.

In its own way, the giant of the diamond world weighed in on the matter of lab stones, too. For every roadblock De Beers has encountered, they've found a solution. Their solution to man-made gems, it seems, is spin. In 2005, a representative told *The Express,* "Our research shows that 94 percent of women prefer the real thing to something grown in a microwave oven in Florida." In 2007, the chairman of African Diamonds, a diamond company connected with De Beers, was quoted in the *Weekend Australian* saying, "If you meet a woman whom you are going to spend the rest of your life with, and have babies with, are you going to give her a diamond made in a lab in Pittsburgh, or are you going to give her the real thing?" And with those three words—"the real thing"—De Beers brushes aside the fact that lab diamonds are physically identical to natural ones. It no longer matters, because the diamond king says it doesn't.

But during the 1950s, even De Beers produced man-made stones of its own. It licensed GE's methodology (for $8 million) and acquired all of ASEA (for an unknown amount). Maybe lab-grown diamonds were starting to rock the boat. As of 2010, two De Beers shareholder representatives, including Nicky Oppenheimer's son Jonathan, sit on the board of directors at Element Six, a company that supplies, among other things, synthetic diamonds. Of course, De Beers' involvement in man-made diamonds is geared toward technological advances—including scientific and medical applications of the gemstone—rather than a man-made consumer market. For that, they've got "the real thing," a phrase even my father uses.

Man-made diamond companies such as Apollo and Gemesis are not looking to fool anyone. They are up front about their product and the fact that it is different from a mined diamond. What concerns the industry is the potential for these diamonds to make their way into the hands of less scrupulous people who might sell them as mined stones. As Martin Rapaport put it in an article, "We are in a very sophisticated long-term technology race. The technology of creation/treatment versus the technology of detection." Yet he believes that man-made diamonds could even boost natural diamond sales. "They're like copies of *Mona Lisa* paintings. They don't stop the *Mona Lisa* from being valuable. I imagine there must be a million copies all over the place of Van Goghs and all these different paintings, but the original painting is probably even more valuable because there *are* all those copies out there." In his mind, there is room for man-made diamonds alongside mined diamonds. It's all about peering into the future and not being afraid of it. "Why should our industry be limited by what comes out of the ground?" What he is more concerned about are treated stones, because they are more difficult to expose.

Although Carnegie is not at the forefront of the synthetic diamond debate—it doesn't aim to sell its product—the Washington laboratory scene stirred anxiety in the diamond world once before. In the 1970s, a Smithsonian Institution researcher named John Gurney published his findings on the affiliation between diamonds and a type of garnet. Wherever these garnets were not present, neither were diamonds. The opposite did not always hold true, but the discovery still had huge implications for the diamond industry. Anyone with this knowledge could hunt for diamonds almost as efficiently as De Beers, thereby posing major competition. Indeed, Gurney's subsequent research led to the Canadian diamond discoveries during the nineties.

But competition isn't Carnegie's goal. Although the lab has been approached by many different kinds of people who want a piece of their technology and research for commercial aims, Dr. Hemley says they will never create diamonds for the purpose of selling. Just for fun, though, Dr. Yan once took a round brilliant pink CVD diamond to a shop on Forty-seventh Street. He was up front about its origins, and after the store's appraiser observed the diamond, he told Dr. Yan that if it were natural, it would be worth thirty thousand dollars per carat. The stone was between half a carat and one carat. As a man-made diamond, Dr. Yan estimates, it probably would have sold for a few thousand dollars per carat. There are no prices for man-made diamonds on Rapaport's list.

The diamond growers at Carnegie have their eyes on more than wholesale return. In the future, they are hoping for three-dimensional single-crystal diamonds. They want to put a seed in the chamber and watch it sprout diamond from all sides. Dr. Yan believes that people will eventually use diamonds the way they now use silicon. While we sit in his office, he complains about his computer being slow and imagines putting a diamond inside of it. A diamond soaks up heat so well that it can be used in electronics to manage their temperatures. It also happens to be a semiconductor, meaning it can be prompted to behave either as a conductor, a material through which electricity passes, or an insulator, a material through which it does not. What this means is that diamonds can facilitate the construction of machines that will make ours look like the room-sized monsters of the 1950s.

We're not the first generation to think of diamonds as a technological building block. During World War II, both the Allied and Axis forces used them to prepare copper wire for fighter planes. In the future, the gems may improve weapon detectors, lasers, and protective gear. They already play significant roles in medicine as tips to dental drills and in hip implants.

At the moment, diamonds aren't mass-produced for either computer or retail use. In fact, in 2005, Gemesis reevaluated its business model and decided that instead of selling cut diamonds to retailers, it would occupy a twentieth-century De Beers–like role and distribute rough diamonds to cutters and designers. But when man-made diamonds are cheap enough to mass-produce, and Dr. Yan believes that diamond growers like Gemesis and Apollo are not far off from this prospect, that's when we may see a revolution.

I try to imagine a world with only synthetic diamonds. No artisanal miners, no drills, no accidental impurities sliding in between crystal layers of carbon from the lower precincts of the planet. The diamond people gather their merchandise from microwaves and presses. They laser their stones to perfection, ship them out, and forget all about the time when it was geology that determined whether you were lucky or damned enough to find diamonds in your part of the world.

Chapter 16

LifeGem

The corporate offices of LifeGem—the company known for creating diamonds out of the ashes and hair of the deceased—are, ironically, full of life. Things have been particularly busy lately, since the firm announced it was going to turn Michael Jackson's hair into diamonds. But the mood is not always this upbeat. Sometimes a sadness falls over the office, when they are catering to some of their usual clients: average people who have lost someone they loved.

"We get choked up around here sometimes," Dean VandenBiesen, one of LifeGem's three founders, admitted to me. "You see people that are devastated and they can't control their emotions and we're on the phone with people who are crying or they're in our office and they're crying, so we see the emotions and that affects us."

Although Dean believes that celebrity diamonds could become a large segment of LifeGem's business, the company has always been geared toward the common mourner. With its services, survivors can wear their deceased mothers, fathers, children, friends, and even pets on their fingers or around their necks. After all, though diamonds predate people by so many aeons, we're made of the same key ingredient: carbon. Essentially, we're just unprocessed diamond dust, only with more impurities.

The man who came up with LifeGem is Dean's younger brother, Rusty. He was four years old when it struck him that he was going to die one day. Dean remembers it clearly. "We were staying at our grandmother's house in Germany, and she's very religious, and he happened to be sleeping in a room where she kept all of her religious things—you know, paintings and a crucifix and things like that—and he was laying in there, trying to go to sleep at night. It just kind of dawned on him what the meaning of all that was, and so he was upset about it, and from that point forward, he didn't wanna sleep in that room by himself anymore, and so it kind of left an impression on all of us, but we didn't talk about it a whole lot throughout our childhood, really."

Rusty kept thinking about it, though. He thought about what would happen to his body after death, and he wasn't happy with, as Dean puts it, the "current options."

In 1997, he came to Dean to discuss this idea he had about turning dead people into diamonds. Dean was an appropriate confidant not just because he was Rusty's brother, but also because he was a metal manufacturer and had studied geology in college. At first, Dean was skeptical. He thought there was something almost funny about making diamonds out of human remains. When Rusty refused to budge, Dean began to wonder if there were others out there like Rusty, whether people who were anxious about meeting their end represented an untapped market.

For years, as the brothers worked toward a company model and U.S. patent, they told no one about their idea except their girlfriends and immediate family, and a close friend of theirs, Greg Herro, whom they enlisted as a partner. Herro was versed in marketing and Web growth, and was useful on the public relations front. In 2001, the three incorporated LifeGem, and the following year, they began selling diamonds made of people. Some of their first clients were the families of 9/11 victims. Others were the families of cancer victims, the pilot of a commercial flight that crashed, and a baby who had died soon after birth.

To promote their product, Rusty, Dean, and Greg attended the National Funeral Directors convention, as well as the Cremation Association of North America convention. Most of LifeGem's customers still find them through funeral homes, more than twenty-five hundred around the world.

During the early months of business, LifeGem was inundated with calls and received four hundred e-mails an hour. Most were from people interested in using their service, though there was some criticism, as well.

What kind? I asked Dean. I was intrigued by the idea of a LifeGem controversy.

"Oh, just someone saying that, you know, they think it's"—he paused for a moment—"sick."

This isn't an uncommon reaction. The media clips LifeGem posts on its website feature Jay Leno calling it creepy, and Kelly Ripa joking about a tennis bracelet for the black widow, a diamond for each of her husbands. Even Dean and Rusty's friends sometimes tease them about their business. "'I better be nice to you or you'll turn me into a diamond.' That kind of a thing," Dean told me.

I couldn't help but find parts of LifeGem funny myself. The idea of making an accessory out of your dead great-aunt, the fact that they felt the need to post an underlined notice in the cremation education section of their website telling clients: "You do NOT need to send the deceased to our location." I imagined people lugging dead bodies to the doorstep of the laboratory and leaving them there with a note: *Three sets of earrings, please.*

But what I began to realize is that, for many people, LifeGem's allure is far more powerful than all the jokes. If diamonds satisfy man's need to incarnate love, LifeGem satisfies his desire for immortality, or just about as close to it as we may get. In a way, it comes close to the mythic powers the ancient world ascribed to the gemstone. The Greeks once believed that diamonds were the tears of the gods.

On the phone, Dean told me the story of a woman who had kept the cremated ashes of her stillborn baby in her home for forty years until she discovered LifeGem. Finally, she knew what she wanted to do with those remains. It's not uncommon for customers, and their relatives' ashes, to have been waiting a long time before stumbling upon LifeGem.

Dean's most memorable client, though, was an eight-year-old girl by the name of Hannah Rowley, a leukemia patient who learned of the option from a friend during her time in the hospital. Some days before she died, Hannah told her mother that she wanted to become a LifeGem. "I still think that the Lord is going to heal me but just in case He takes me home I want to be a diamond for you Mom," quotes her mother's account on the website. I'm moved by the fact that, faced with the eventuality of her own death, the girl thought to prepare her mother in such concrete terms. And this, I believe, is the simple reason for LifeGem's success: it creates something beautiful out of the ugly business of death.

This is why LifeGem retailed about five million dollars' worth of human diamonds in 2009, why their diamonds sell not only in America but all over the world. They are particularly popular in Japan and the UK, but reasonably so in Taiwan, Belgium, the Netherlands, Luxembourg, and Germany—all countries with high income levels and cremation rates. As another mother put it, on the German LifeGem website, "My dead son now radiates at me as a diamond."

The LifeGem process resembles a cross between diamond synthesis, GIA certification, and necromantic alchemy. The company is so protective of its innovation that I wasn't allowed to visit, but I got a detailed rundown from Dean, who handles the scientific parts of the business. When a person signs up, he or she receives a plastic container that says "LifeGem" on it, tamper-proof tape, and a cluster of papers to sign. Although some people like to personally drive their loved one's remains over to LifeGem, most send them by registered mail.

As with diamonds at the GIA, LifeGem makes certain that an ID number accompanies a person's burnt hair or remains wherever they go. In the first step of the procedure, which they call "carbon capture," an eight-ounce cup of ashes is placed into a "vacuum induction furnace," a machine that sucks the oxygen out of the remains, thereby safeguarding the carbon from decomposition. This reduces the eight-ounce bulk (almost 227 grams) to about two grams of nearly pure carbon.

Next, the remains are transferred to a metal "crucible," which looks a bit like a salt shaker one might find in a cafeteria kitchen. The crucible is engraved with the remains' ID. Ashes inside, it is cooked to approximately three thousand degrees Celsius under "special conditions" that Dean would not reveal. Under these conditions, the carbon turns to graphite. They call this step purification. I asked Dean what purified carbon looks like.

"It's a sparkly black powder." Just like diamond dust.

Next, the graphite is subjected to high-pressure, high-temperature conditions in a diamond press, a beige-colored machine. Inside is a metallic pyramid mounted atop a circle. The environs inside the press separate the carbon atoms in the graphite and allow them to reengage in shared electron bonds to form a diamond. Unlike Carnegie, LifeGem doesn't start with a natural diamond; it uses HPHT to grow its gems, not CVD.

After a rough stone emerges from the press, it is subjected to the same cutting and polishing as any other diamond. It is inscribed with a lasered ID on its girdle, just as the GIA offers. It is even graded by a resident GIA-trained gemologist according to the Four Cs.

Although Dean wouldn't tell me exactly how they launder their wheels between diamonds to make sure the residue of one person doesn't make its way into the diamond of another—"proprietary information"—all in all LifeGem creation is a very clean procedure. Unlike death, it even has a return policy. But then, there are little reminders that the ingredients of this diamond once belonged to a human body, perhaps the most visceral of which is the fact that, although LifeGem diamonds come in colorless, blue, green, red, and yellow, the company does not—*cannot*—guarantee the shade within each color group, because each body's remains contain unique traces of elements such as calcium and boron that affect the hue of the diamond they become. Every diamond, even a natural one, is tinted by the impurities trapped inside of it. Only here, the calcium or the boron used to belong to a body, and the presence of these elements could be the result of a person's environment or nutrition, though it would be difficult to track any specifics.

For similar reasons, LifeGem makes no promises as to their diamonds' clarities and imperfections. They've never made a stone that's clearer than VVS (LifeGem clarities typically range from VVS to I). To explain this policy, the company avails itself of a patron's quote: "As one of our clients said about her husband, 'He was perfect, yet certainly not flawless. I wouldn't expect his LifeGem to be without flaws either.' At the LifeGem state-of-the-art diamond facility, we pride ourselves on creating the highest quality diamonds possible, but please understand that this is a scientific process that produces a unique diamond every time."

Exact carat weight is also slightly variable, though this has more to do with the imprecision of current technology than the uniqueness of human bodies. When you order a LifeGem, you can choose a diamond that weighs between .20 and .29 or .30 and .39, etc., but not an exact-quarter carat stone. The largest diamond they offer is two carats or slightly larger. What their scientists do know is that the longer graphite spends in the press, the larger the emerging stone will be.

Cut is the only C they'll guarantee, and they offer pretty much any shape diamond. Round, radiant, princess, and heart are their most requested. Prices vary depending on color and carat weight. Colorless and blue are the most popular and also the most expensive. A .50- to .59-carat diamond in this category costs $7,899, or $7,299 if you are ordering more than one. A red or green stone of the same weight runs $7,299 and a yellow, $5,999. LifeGem offers payment plans, including, of late, the "plan ahead" option, evocative of life insurance.

If the customer wishes, a jeweler will set the diamond in a ring or a pendant, just like the setter set my father's diamonds. I was expecting the jewelry to be very tacky, but I was wrong. The LifeGem online catalog offers a collection of almost antique-looking options. There are even a few simple "gents" rings, though Dean told me most of LifeGem's customers are women, "which makes sense, because I think women are more comfortable with diamonds."

As with regular diamond sales, which peak in early winter because of Christmas, LifeGem gets a lot of its orders in the colder months, when people are at their most physically vulnerable.

The majority of customers choose the cremation option, even though LifeGem has the ability to create diamonds from the deceased's lock of hair. They require only the quantity removed during an average men's haircut.

"Maybe," Dean suggested by way of explanation, "they feel that that's more the whole part of the person, rather than just a part of the person."

His most outlandish request involved a violin. "It was an artist who burned a violin and sent us the ashes from that burnt violin to have a diamond created, and then I think they were gonna use that diamond from the ashes of that violin to be used as a needle in a record player to play back music that had been played on that violin." Dean laughed.

"Oh, how meta," I said.

That was the only inanimate object they've been paid to make a diamond out of. Now they're working on a project commissioned by the Royal Arts Academy in London to turn polar bear ashes into a half-carat colorless diamond for a display.

And then, of course, there's Michael Jackson's hair. The story of how Rusty and Dean VandenBiesen came to acquire the lock begins with another musician.

"Back in 2007, we had the idea that we wanted to try to get publicity about our capability to create diamonds from locks of hair, so we were thinking, 'Well, what's the best way to make this public announcement that we have this capability?'" They decided the answer was to create a celebrity diamond, and discovered a man by the name of John Reznikoff who had a business collecting famous people's hair. As it turned out, Reznikoff was in possession of a lock of Beethoven's, so LifeGem joined forces with him, turned Beethoven into a diamond, and sold it on eBay for more than $200,000 to a private buyer. They split the profits.

When Michael Jackson's death was announced, Rusty and Dean wondered if they could do the same with him. So they called Reznikoff to see if the collector had managed to get hold of any Jackson hair. He hadn't. Two weeks later, he called them back. The executive producer of Jackson's famed Pepsi commercial, in which Jackson's hair caught on fire, had contacted Reznikoff. This producer had doused the flames on Jackson with his Armani jacket and kept a lock of his hair. The producer sold Jackson's scorched hair to Reznikoff. LifeGem then purchased some of the hair from Reznikoff, and announced their plans to make Michael Jackson diamonds. Dean e-mailed me a photograph. In it was a picture of Jackson and the producer, as well as a pitch-black stringy collection of hair atop a DNA verification slip from Cedars-Sinai Medical Center lying on the inside of a green-gray sports jacket, the Armani label prominently displayed.

But the least-known success of the product is that Life-Gem has allayed the anxieties of its inventor. I wanted to know if Rusty felt better about death since he created the company. Dean said, "Yeah, this was something that made him feel a lot better about mortality. You know, one of the things about Rusty is that he likes to be around people. Doesn't like to be alone, likes to talk to people. He's one of those people that'll spend a lot of time on the phone if he's by himself, and so just the thought of, say, his close friends or family having a diamond made from him after he's gone kinda gave him this feeling of, 'Okay, I won't be forgotten. I won't be alone,' so that aspect, that thought, made him feel a lot better about mortality."

A few days after speaking with Dean, I went to see Brian McKee, a mortician at Thibadeau Mortuary Service in Maryland. At the time, Thibadeau shared a building with a second funeral home in a semi-suburban part of Silver Spring that's overwhelmed with automotive businesses. The other funeral home's navy Cadillac hearse was parked in the driveway I had to pass through to enter Thibadeau. By the door was a bouquet of umbrellas whose handles were labeled with Thibadeau's information. A rectangular military flag frame stood on a cabinet facing the door.

Brian led me to a side room with a glass table and a modest display of memorial and funereal products. Two cases housed black and white plastic urn vaults, a sign-in packet, ash dispersal kits, and, on one of the bottom shelves, a baby casket with a pillow inside. Farther into the office was another set of shelves displaying all sorts of urns, including The Presidential Urn (a large black wooden box), The Colonial Urn (a smaller reddish one), a few marble urns, a companion urn (which holds two people's ashes, side by side), and a metal heart-shaped urn. Then there are the keepsake products—glass domes with jewels inside, framed psalms, even a teddy bear. On the table lay a personalized Thibadeau card with a heart-shaped piece of biodegradable paper—seed cards for gardening. For my benefit, Brian had also set out two LifeGem brochures on the glass table.

Brian is married to a funeral director from another mortuary service and is surprisingly funny. You really have to have a sense of humor when you spend your days staring down death. You also have to be open-minded when you work in the interfaith funeral business, because the range of people's wishes when it comes to the remains of their family members is unimaginable. There are those who pay to have their loved one's ashes propelled into space.

Brian doesn't pitch LifeGem to all his customers when he reviews his list of services, but if a client is choosing cremation, he may ask, "Are you interested in any kind of keepsake items?" and if the client seems to be the type to appreciate it, he will mention LifeGem alongside the picture frames, the glass domes, and the crystals grown from human remains.

For making the liaison and packaging the ashes, the funeral home receives just over ten percent in commission. They get their cut after LifeGem receives the second half of their fee from the client (customers pay LifeGem half the cost up front and the rest when their diamond has been created). Thibadeau has sold two LifeGems so far, one to the wife of an American soldier who committed suicide and one to a man whose wife died of cancer.

When I told my father about LifeGem, he said he found it a bit disgusting. "I think you should let someone rest in peace and not carry him around like a necklace, a token, or something like that. Like a lucky charm." Later, he added, "I hope you guys don't do that with me."

Chapter 17

Safe Crumbs

The day my uncle died, New York City went dark. We pointed this fact out to my grandmother on our way to the funeral the next day, August 15, 2003, as if to say, "They turned out the lights for Steve." There were blackouts in a handful of other northeastern states, as well. The entire region seemed to be mourning him.

Steve had been alone when he found out he was dying. In the late fall, he'd gone to the doctor in Toronto because of leg pains. He was given gout medication, but the pain persisted. A second doctor found blood clots in his legs, then blood clots in his lungs, and then masses in his lungs. My father had a doctor's appointment of his own the day Steve was scheduled to see the oncologist, but Steve told him to come straight to the hospital instead. By the time he arrived, Steve had already learned he was terminal. He didn't cry when they told him the news, and he didn't cry when he passed it on to my father.

Together, they went back to the apartment my uncle inhabited alone, with his watches. The watch parts were everywhere. In the office, they dwelled on tables and chairs and windowsills. They poured out into the living/dining area, and lay around like witnesses as Steve grew fragile and anemic over the course of a few months. My father suggested that he help Steve put the pieces away, but my uncle wouldn't let him touch them. He kept saying he wanted to buy time, to get back to work.

"Look," my father would say, "you can't work on it now. You're feeling too weak. Let me get it out of the way so you have more room for yourself."

Before my father told us about the cancer, Steve took one last trip to New York. He could still hide it back then. We shot a series of family photographs in our kitchen, some of which are framed in the office, near my father's desk. Sometimes I scour my few memories of that visit for clues that my uncle was standing face-to-face with death. All I remember is that I was having a tough time and that he put a hand on my sweatshirted back and said, "Hang in there." It was the most intimate thing he'd ever said to me, and I did. Shortly thereafter Steve was back in Canada, and my father was pulling over on a street corner one night after picking me up from Penn Station—I was home from college, visiting—to tell me the news.

By the end of July, Steve had yielded on keeping the watch parts out. My father set about rounding up the constellations of timepieces that had taken up residence around his brother's home. Carefully, he laid the delicate parts into trays and boxes. He didn't know exactly where everything belonged, and Steve was too sick to give him instructions, so he went by eye. "You know, when he had little projects, I tried to keep 'em together and put 'em together as best as I could, but I think I did a pretty good job."

My uncle spent his last few weeks in my grandparents' apartment—one last time, my father had persuaded him to come to New York. Our refrigerator became stocked with Paul Newman's brand of pink lemonade, because it was one of the few things that wasn't repulsive to Steve's taste buds, warped from chemotherapy. I watched as my uncle shrank on the convert-a-couch in Oma and Opa's TV room. In relation to him, the bed was a vast expanse.

On August 11, Steve's breathing became labored and heavy. My father called Hatzolah, the Jewish ambulance service, and checked him into the hospital. Three days later, in a room at Columbia Presbyterian, Steve died in his arms.

"I covered him and stayed with him till the morning, till I could reach the funeral home." He wouldn't let them take the body to the hospital morgue. He waited until Steve had been brought to the funeral parlor the next day, and then made his way to his parents' home.

Soon, the mirrors in Oma Dora and Opa Yankel's apartment were covered. We taped sheets of leftover wallpaper to every mirror. Once, though, during the seven days of shiva, I spied my grandmother peeling back the edge of the wallpaper in the bathroom to take a peek at herself. Other times, she cried out for her mother, *"Maminyu, Maminyu,"* as though asking to be taken back to girlhood, to a time before any of these absurdities were possible—surviving a war only to lose a firstborn son to cell growth. I took the blackout as a sign of the universe's responding to our suffering.

The shiva filled my grandparents' home with people. Among them were a number of Forty-seventh Streeters who came to pay their respects. The Club had hung a sign on one of its bulletin boards telling the dealers that Jack and Paul Oltuski had lost a son and brother.

When a member of the Diamond Dealers Club or his close family dies, the Club puts up a notice. Members pay an annual fee to the Benevolent Fund so that it can help allay the cost of burial when the time comes. But while many Forty-seventh Streeters consider themselves part of a diamond-dealing family, "mourning is a very personal thing," Eli Haas, the former president, told me, so other than some financial help, the Club leaves the mourners to grieve in their own way.

Most of Opa Yankel's diamond district friends who visited were Hasidic. They approached my grandparents and father in the front of the living room, where they sat on low chairs designed especially for mourning, and quoted them the traditional Hebrew words of sympathy: "May the Almighty comfort you amongst the mourners of Zion and Jerusalem."

One of the visitors was the son of the diamond manufacturer from whom my father and Steve had rented their first office on Forty-seventh Street, where diamond dust had flowed through the air courtesy of the ventilation system. Several of my father's district friends also came to the funeral in Queens. A dealer with a booth in 10 West was a pallbearer.

Before Steve's death, Oma Dora and Opa Yankel had been spending part of the year in Germany, the rest in New York. Now they gave up the lease on their Frankfurt apartment. It was foolish, they thought, to be so far away from family. Steve's death brought my grandparents to America for good. A few weeks after the funeral, my father returned to the district. Shortly after, my grandfather rejoined him. The street swallowed them up, allowed them to come to work each day as new men. When they stepped onto Forty-seventh, it was as though they were stepping into a different, more radiant, world. Only now, when he came home, my grandfather spent a few minutes each day sitting before a full-spectrum lamp my father had bought for him and Oma. Sometimes, without warning, tears would illuminate his eyes at the Shabbat table, and my father would bring him a shot of vodka. It was our tragedy, and it had become part of our bequest just as much as the villages we stemmed from and the war and the diamonds.

In the fall, my father posted a message to Steve's devoted legion of clients on Northern Time's website. "Dear friends and valued customers, I am saddened to inform you that my brother Steven Oltuski has passed away. In December of last year, Steven found out that he had a terminal case of lung cancer. Despite this awful discovery, he was determined to fight as hard and as long as he could, to be able to win as much time as possible and hopefully be able to continue his life's work and enjoyment—Northern Time." He'd asked me to edit it, as he did all his most formal documents, such as letters to boards or the government.

In his home in Toronto, Steve has all the time he wants. My father kept the apartment as a sort of shrine, hairsprings and all. The floor-to-ceiling unit stands untouched. In it lie the glass casings, the leather bands, the dials, and the tiny cogs that Steve was saving for future timepieces he thought might come his way. And tucked away in a safe-deposit box at a nearby bank in Toronto are the watches, some still ticking, others waiting to be reawakened.

Five years after Steve's death, my father and I are standing in the back of Jimmy Buyout's gallery in New Jersey. The gallery is filled with framed paintings—Jimmy buys even more art than jewelry, and he owns one of the largest collections of Pennsylvanian impressionists in the country and has written a book on the subject.

My father and I have driven here together for the day to look at a new jewelry estate Jimmy has purchased. First we chew the fat.

"Where's your car?" my father asks. Although Jimmy owns an almost unfathomable amount of jewelry, what impresses my father most of all is his sports car. On a small partition near his desk, Jimmy keeps a miniature model of the Ford convertible that he owns.

"At home," says Jimmy.

"I love your car," my father murmurs boyishly.

Cars are important to my family. In the forty years my grandparents lived in Germany, they never bought a house, but they did keep a brown Mercedes coupe parked in their rented garage spot—their ride. And as soon as they could afford it, my parents got their own Mercedes in New York.

At Jimmy's, my father rolls the convertible around on top of the partition. Then he surprises me. He tells Jimmy that he has a collection for him to buy out: Steve's. He says that his late brother just bought and bought and bought. He tells Jimmy that Steve had all kinds of things, like hairsprings and other watch parts.

"So how much you want for it?" Jimmy asks.

And just like that, everything—a whole lifetime of gathering—comes down to this one question. But for jewelers, the difference between a private collection and a work collection is immense. My father isn't selling Steve's art or his furniture. He's selling his product. Dealers aren't sentimental about jewelry the way other people are.

Without any ado, my father states his price. "There's vintage lighters," he adds.

Doing business with Jimmy is a different way of buying and selling jewelry. He leads us through a supply area and up a carpeted staircase. This is the life of a dealer: being shown to messy back spaces where all the real treasure resides. At the top of the stairs, Jimmy types a code into a number pad on a door, and we enter what looks like an attic filled with every kind of belonging imaginable. Paintings in gold-colored frames, a flat-screen TV on an antique table, a large signed boxing glove (one of Jimmy's great regrets is not having competed in the Golden Gloves), a tin statue of a boy, a hat placed on top of the boy's head, suitcases, an oversized stuffed Gumby, and a bottle that says "Purple Magic" filled with violet liquid sitting atop an aged table.

My father and I take a seat at a large wooden dining room table, and Jimmy carries out trays of jewelry. My father sorts through them like an archaeologist, sometimes bringing a loupe to his eye. On the table, he has set up a miniature digital scale and a navy blue toiletries case labelled Delta Business. The old Pan American case was retired a few years ago.

Jimmy tells me about how he first started picking through people's trash and selling it at flea markets and garage sales with his mother at the age of twelve or thirteen. He turns the television on. These days, he's been keeping track of the value of gold and silver and purchasing it on the Internet. Prices have dropped tremendously, and Jimmy is banking on people's taking their money out of the banks and investing in precious metals.

Eventually, he gets back to the story. At one of the markets, he met a seventy-year-old ex–pro golfer named Chick who owned a coveted booth at a flea market Jimmy compares to 10 West Forty-seventh Street. When he was thirteen, Jimmy chipped in on rent and joined Chick's booth. Once a week, at four thirty in the morning, Chick would come pick Jimmy up at his parents' house and take him to the flea market, where he would sell. Then he'd drop him off at school. "Mondays, I was always late for school."

Jimmy hasn't stopped buying since he started. Procuring garage sale leftovers is not so unlike buying off showcases in Vegas. Only the profits (Jimmy's margins tend to be ten to twenty percent) and the price tags are different. The technique, he has honed over a lifetime. Although people speculate about what goes through his mind when he buys, Jimmy says it's not mystical at all. "I just kind of look at it and I can add it up quickly. I can add this whole pile up in two minutes. I'm kind of like a Rain Man."

All this buying translates into an astonishing amount of objects. "My biggest fear," he told me that day, "is being one of those guys where, when they die, they have one of these two-week-long auctions, because I've been buying those guys my whole life, and I always say, 'Why do these guys save all this stuff? What was wrong with these people?'" and at this point my father chimes in, "It was a mistake," laughing.

As we sit at his large table in his atticlike room, Jimmy tells my father and me about the time he once bought $2,011,000 worth of diamonds from a dealer in approximately ten minutes without even opening the display case.

My father wants to know how Jimmy can be sure that the owner of the merchandise was being honest about his pieces. I can tell he is slightly dumbfounded by this system. Jimmy seldom touches the merchandise in question or uses his loupe to view it.

"Well, I ask a lot of questions. I can ask you a hundred questions and know every answer you told me." He'll phrase these questions in different ways, grilling vendors like the CIA to arrive at the true value of a collection. But he doesn't need a lie detector; the lies and errors come out all on their own.

He'll ask, "How much for everything?" He'll ask how much has been offered, what something is worth, how much the vendor would buy the items back for, who else has made an offer, what was the lowest number he's ever quoted. There are different genres of questions, but they wind their way to two main pieces of information: How much is the showcase worth? And does the seller know how much the showcase is worth? "I need to buy from people who don't know exactly what they have."

Some of the questions he asks jewelers are for their benefit as well as his own. "I don't want to put anybody out of business, so the first thing that I'll say is, 'Do you own everything?'" Jimmy doesn't buy memo pieces, because he assumes the jewelers couldn't afford to buy them outright, which means they certainly can't afford to sell them to him at a discount.

By the time he has asked a couple of questions, Jimmy is in possession of all the answers, and it's only the merchant's inconsistencies that matter. In his attic, which he doesn't call an attic but a boardroom, he gives me a two-million-dollar example. "He had a pair of fifteen-carat browns. They were EGL-J-SI-other thing," he listed their specs. "And I said, 'What do you ask for them?' He said, 'Six forty.' And then I asked a bunch more questions, like, 'So what are they worth? What have you been offered? What would you buy them back for? What are you asking for them?' Or, 'You're asking six ten for them?' And if he says yeah, then I know he's not forthright."

Little of Jimmy's system comes from formal training. "I haven't ever read a book that I can remember. The only book that I read cover to cover is the one I wrote, just to proofread it." Jimmy's college career lasted one semester, and he only waited that long for the boxing, which he still keeps up at age forty-five. (My father was recently forced to give up sparring when he took a knuckle to the eye.) "I knew after a few days that I wasn't going to continue, so I just stayed throughout the semester mark period, because I was in the boxing gym the whole time, but then when the grades came out, I got a zero zero, so I couldn't stay." I feel bad, but Jimmy is laughing. "But I didn't want to stay. In my mind, college is to teach people how to make a living. I know how to make a living."

His greatest asset, though—his memory—he was born with. "I already don't know your name," he says to me, unapologetically. "Somebody could tell me their name and two seconds later I can't remember it. But if you were a dead artist and I once had a painting by you, I'd remember everything."

When he is finished browsing, my father asks, "More?" but Jimmy tells him that was all the jewelry he had to show him.

"What was this?" I ask. "From where?"

"Chicago," says Jimmy. My father says he knows the owners, "and they must have sold you not their . . ." He trails off to indicate *not their best merchandise.*

"No, I bought everything except he had one safe full of—"

"Nice stuff," my father fills in. He is not trying to make Jimmy feel bad. He's just not the type to keep opinions to himself unless they relate to his own dealings. In this way he is like Opa Yankel.

"Pretty nice stuff but this is most of their stuff. I sold some of it, but you could see, it's a lot of stuff still. In his safe was mostly expensive watches, big stones. I think he's got some in safe-deposit boxes."

My father perks up. "Oh yeah, he has big stones?" he asks, practically licking his lips.

"Not a lot." Jimmy says that, supposedly, he'd bought out everything the estate owner had in Chicago.

"Oh, okay. Yeah."

"I bought out four safes."

"That's a lot," my father concedes.

Finally, they turn to prices. Here, there are no negotiations. Jimmy cites only one number, and my father either takes it or leaves it.

Jimmy gets down on his knees next to the table and tosses a big ring onto it, as though the jewel is made of plastic. "Twenty-seven fifty." When he utters the numbers, his voice takes on the decisive tone actors use in movies to announce final offers. "Thousand dollars." He holds another item in his hand, drops it on the table, and flicks it. "Seventeen hundred."

When you've dealt with big-ticket items as Jimmy has, five-thousand-dollar pieces begin to feel cheap. But even the expensive purchases only thrill him temporarily. "I guess a good analogy is like I'm a dog chasing a car. You ever see a dog chase a car?" he asks me. "Once you catch it, you don't know what to do with it." Buying is Jimmy's high. After that, it's only a matter of business.

"What'd you say this was?" my father asks. It is the same question Jimmy asks his buyers. Perhaps he is learning some technique, but probably he has just forgotten. Either way, Jimmy doesn't blink. "Seventeen fifty."

When I ask how he figures out his prices, my father cuts in to joke, "He makes them up." In reality, Jimmy relies on other people's price tags, adding them up swiftly and then adjusting for profit margins.

When he's all through, my father says, "Jim, sorry there's not a lot there, this time." He's chosen only two items, an aquamarine ring and a brooch.

"Less than a billion?" Jimmy asks. He's not insulted, he just wants numbers. But my father doesn't answer.

Before we leave the room, he asks if Jimmy is interested in Steve's watches.

"Yeah. I mean, is there money left in it?"

"I think so."

Jimmy starts up with his deluge of questions: Who looked at it? How many finished watches are in the collection? Did they make an offer?

"He was like you," my father says. "He had a watchmaker case, and he just filled the drawers up with watches and projects and what-have-you. Then he had—"

Jimmy interrupts to ask if everything in Steve's apartment is for sale. My father says that only the watch parts are, not the furniture. He hasn't changed anything about Steve's home.

"So what are you gonna do with it?"

"I'm going to keep it. I'm going to keep it, 'cause, you know, it's memories, you know, I was very close to him. I didn't touch anything in there."

I imagine Jimmy wandering through my uncle's apartment, eyeing the hundreds of watch parts and trying to gauge their worth. Although in our grief, the selling of Steve's timepieces feels like something uniquely devastating, almost all estate jewelry once belonged to someone who has died. Our sorrow is nothing special, and loss is what gives the world its antiques.

As for our family, life doesn't go on, but it also does. I had never seen my father cry before Steve died, and then suddenly it was a common, unspectacular occurrence for his eyes to get so red they contrasted with the green in his irises. I don't remember tears ever coming out, just his eyes filling, as though there were a secret veil, an invisible loupe that kept them lodged bravely in his eyes. My grandparents more or less stopped dancing at weddings and bar mitzvahs, but sometimes they still go out to card clubs with their friends and even allow themselves to win some money. Oma gets her hair done. Opa gets a new La-Z-Boy chair to watch the news in. He calls it a Crazy Boy. He stays in shape, swimming and stretching himself into oblivion.

My father and Jimmy talk about getting Jimmy to Toronto to look at the collection. Lately, he has been hiring private jets. My father doesn't know much about the logistics of traveling by private jet. But because Jimmy is a man who treats the people he buys out as equals, he can pull off his next line without seeming snobby: "Well, that's how I roll."

Chapter 18

I Want You to Take Care of Your Daughter, and Other Auction House Follies

In the fall of 2008, the banks begin to topple. They go like sand castles at high tide. A few blocks away from my father's office, the newly minted Bear Stearns building becomes the JP Morgan Chase building. Soon, we watch Lehman Brothers go under, Merrill Lynch sell itself. On Forty-seventh Street, the mood is gloomy, but it is auction season again. So while the rest of the country reconsiders its purchases—holding off on houses, expensive cars, and brand-name goods—the jewelers are buying as much as they can. A rash of sales fills the months of September and October. In addition to the two largest houses, Christie's and Sotheby's, multitudes of midsized and smaller houses, including Doyle's and Bonham's, lay out their jewels for dealers and privates. I'd always imagined auctions as fast-paced, frenzied circuses with people shouting out numbers, scratching their heads, pulling their ears, and thumbing their noses. Auction day is indeed fast-paced, but the process leading up to it is lengthy, elaborate.

It begins when a catalog arrives in the mail. Dealers thumb through it, looking for pieces that catch their eye. Then, a week or two before the auction, they are invited to come preview the goods. This is when they scrutinize the jewelry and diamonds up close, deciding what they want to bid on and for how much. Sometimes, while my father is at a viewing, the department heads will give him a sneak preview on an upcoming auction's highlights. "Take a look at this," they'll say, handing him a gem of tremendous clarity and size. "They'll sort of tease me with a piece."

After Jimmy's, my father and I continue on to Washington, D.C., to look at the contents of an upcoming sale at a small family-owned auction house. In the car, I ask him what happens to the merchandise of a dealer after he retires.

"Meaning if he doesn't have relatives to continue his business?" I find myself wishing he didn't spell it out. It makes me sad that there will be no son or daughter to take over. This is, after all, a business of sons and daughters.

"Yeah," I say.

"Well, either he sells his business to one person, his entire business, or it gets auctioned off."

"What do you think you'll do?"

"I have no idea. I'm not planning on retiring."

I ask him if he's seen friends go through liquidation.

"No. Actually, a lot of people in our business on Forty-seventh Street, they work till they're carried out horizontally." He laughs, so I do, but it's not funny to me. My father changes the subject to point out a truck full of ponies on the road. A few weeks ago, one of his earliest diamond friends on Forty-seventh Street went into the hospital with a heart condition.

When my father first started working in the diamond district, he did a lot of business with older dealers, his father's friends. "Now a lot of the old-timers died out and a lot of the young people came in, I guess *I* might be considered basically almost an old-timer now," he said to me.

Just yesterday, the son of a colleague came to my father's office to show him some merchandise.

"So you're the young Mr. Danzig." The young Mr. Danzig (a pseudonym) was twenty-eight, still a boy in this trade. My father sat down with him and let him peddle a few rocks.

"What are you asking for the stones?" he inquired.

Danzig stated his price.

"And what do you really want for this stone?"

They talked a little bit. My father asked Danzig if he knew what the stone's list price was. He looked at the list. He told Danzig that he knew his grandfather. The boy said he himself didn't know his grandfather.

My father took one of Danzig's stones and examined it under his white diamond lamp, harbored in the valley of a folded piece of colored paper. He brought the diamond to his mouth and breathed on it, then held it under the lamp once again.

At one point, he told Danzig, "The truth of the matter is, I hate to buy I-colors." He asked Danzig if he knew why. Danzig asked why. He had to. My father in turn asked if Danzig knew the list. "It's not a white stone and it's not a cheap stone," he explained. A white stone is expensive but easy to sell. A cheap stone is easy to buy. An I-color is neither.

The boy said he had Js, rounds. My father said he was into cushions. Soon, they moved on to a different diamond.

"I'm not gonna make you an offer on this," my father said frankly, "because to tell you the truth, I wanna steal such a stone." By steal, he meant the low thirty thousands.

Before Danzig left, he told my father, "I'll learn what you like, and I'll try and get you first crack at it . . ." I watched as he made his nervous and deferential exit from the office.

My father and I arrive at the auction house in Washington an hour late. The man showing him the merchandise, one of the house owners, asks if he needs a watch, but nobody laughs. My father blames traffic. The owner takes us up to the auction room, which is set up for an event, complete with a podium. The room itself looks like an antique item, with old wooden floors and large windows. Since it is a small house, its dealer previews are private.

We sit down at a table in the back next to the display cases. My father puts on a special loupe that attaches to his glasses with a clip, and starts to sift through the merchandise. The stones leave a little indent on his finger. He moves one of his sample diamonds onto the base of the diamond light the house has provided him with. Then he picks it up with a bronze contraption that looks like a pen. When he presses down on the instrument's top it releases metal tentacles as thin as spiders' legs to grasp the stone. He uses the spiders' legs to hold the loose stone next to a diamond necklace to compare their color. Soon, he measures the carats of the necklace's diamonds. I hear him mumbling quietly, reciting a private litany of lot numbers and values. His left hand shakes a little bit as he gently grips the jewels.

These quiet moments of the business are my favorite, my

father hunching over the jewels to search them for flaws, both on their exteriors and deep within their interiors. In Hebrew school, they taught us a biblical line that goes, "The hidden ones are for God and the apparent ones are for us and for our generations." With his loupe in his eye, his forehead pressed so closely against the diamond lamp that it cowlicks the front strands of his hair, my father looks for them all.

His attention to detail is old and familiar to me. When I was a kid, he used the same scrutiny to check me for ticks when I'd gone out in the country in the summer. He bought a dermatologist's set of magnifying glasses, similar to the loupe, and investigated my scalp, my arms, my legs. When I was even younger, after I'd had my evening bath and blow-dried my hair, he would run his fingers through my curls to check for damp spots that could lead to a cold.

But all those years, I was watching him, too. I knew the patterns on the suit jackets he liked to pair with jeans for workdays and sometimes also weekends. I knew intimately his neatly trimmed fingernails, the lack of wedding ring on his fourth finger (both my parents lost this piece of jewelry long ago), and the small flakes of dry skin that separated and nestled within the hairs of his eyebrows, brought on by a harmless dermatological condition.

When my father is finished looking at the collection of rings he attempts a sale. According to him, it never hurts to try. So he asks the owner when the next auction is. "Want some goods?"

"Nah," says the owner, "I got a hundred lots already."

Before we leave, my father will make a second attempt, saying, "Well, let me know if you're running short on goods. I'll send you some well-priced things."

The man chuckles.

"Yeah," sings my father in a playful voice, meant to persuade him to consider the proposal.

In the car, he tells me he thought the trip would be more fruitful. The items looked nicer to him in the catalog than they did in real life. "But," he says, "you have to come and see. Veni vidi vici."

As we pull over in front of my building in Maryland, my father tells me how hard it is to make a living these days. For every item he sells, there is a bill waiting at home.

"I'm like Sisyphus," he says, and he is right. His life revolves around a rock.

A few weeks later we are at Sotheby's, the day before their Important Jewels auction. It is my father's second visit in two weeks. The first was an official dealer's preview. Today, the public gets to look at the auction goods in the gallery.

Sotheby's resides in a large and formidable building on the Upper East Side of Manhattan. The structure's facade is almost all glass, giving the illusion of transparency. The day of the preview, dealers arrive like suitors, dressed up and waiting for a chance to spend time with the jewels. My father wears a shimmering gray suit jacket and carries a leather briefcase; his black leather U.S. Open shoulder bag is at home. He has put on the ritz without looking as if he tried too hard, and I find a kernel of pride blooming in me. When I compliment the briefcase, he admits that its screws keep coming out. "One time I was walking in Paris, and it just went plop," he says loudly in the elevator, and I wonder if this is something he should be advertising.

Upstairs in the jewelry department everyone knows him, smiles at him; he is in his element. As we register, he says to a passing employee, "I brought you chocolate." She hesitates, knowing he is a joker, but then stops, trusting. My father reaches under his jacket and pulls out a Slim-Fast bar.

"Oh." She waves him off.

"Now you're picky?" he asks.

We turn into a special dealer's room. It looks a little like a small European café. People sit opposite each other at elegant tables with beige leather tops. There is coffee and seltzer, and jewels are set out in velvet-lined trays. Dealers mark off the pieces they want to view on a checklist of items, and Sotheby's employees deliver them to the tables. On each table stands a diamond lamp.

Most of the women in the room are temporary employees Sotheby's has hired for the auction season. All of the dealers are men. My father and I are seated at a table in the back with a microscope and professional fluorescent light. This is not the little flashlight-sized affair my father owns but an entire chamber in which to place the stone, and goggles to look through. My father checks off items on the lot list, and while we wait for them, he chats up one of the seasonal helpers, poking fun at her German accent.

This time around, he notices a spoil in the enamel of a brooch he'd overlooked last week. It's easy to miss. The piece is composed mostly of a hunk of carved ruby, which sits atop a bush of emeralds and diamonds. The black enamel is almost hidden. It forms a thin border around the emerald and crowns a diamond at the top of the brooch. Technically, the enamel is fixable, but that's not good enough for an antiques dealer. Sometimes fixing jewelry is worse than leaving it damaged.

While my father looks through the tray on his table, a woman comes over and asks to borrow a piece.

"For whom, for Moses? No, no, no, no, no. No, no, no, no." His reaction tells me that the old dealer sitting at the table behind ours is important.

Later, when another item is picked from his tray, he gets serious. He tells the staff they have to keep track of what they take from him. He doesn't want a piece to disappear on his watch.

My father wears his loupe around his neck on a blue band. He studies the jewels, often pivoting them back and forth in his fingers to measure their shine, their life. Quietly, he says to me about one of the diamonds that it is a very good SI2 (Small Inclusion 2), that its flaws are well spread out and don't jar the eye.

He takes out a smaller loupe that magnifies the diamond sixteen times instead of the standard ten. He's surprised the GIA gave the stone an SI2 rating. He speaks softly about the diamond so that no one else can hear his admiration of it. As he loupes it, I can see his eyeballs moving, his eyelids widening, as though they are going to swallow up the stone. He examines the diamond again with his regular loupe. Then he has me take a look. I spot some black matter, some white blurs. They are fairly obvious, as far as diamond imperfections go, but this is the first time I am easily able to identify the minuscule properties that make or break a diamond's worth.

Gary Schuler, the director of the jewelry department, has stopped by our table to chat, to take care of good customers. My father is telling a story about a wedding band he bought at a Sotheby's arcade sale. The day he was supposed to pick it up, he got a call from an employee who told him the ring wasn't ready; one of the salesladies had tried it on to show a customer and couldn't get it off. Diamonds surrounded the band, so there was no way to cut through it without cracking the stones. My father called Sotheby's again the next day. The ring was still stuck. The saleslady's finger was swelling. My father was leaving for Germany the following day. Gary interrupts the story to make fun of him for not caring about the saleslady's finger and just wanting his ring.

Finally, my father tells us, they took her to the doctor, who tied floss around the ring, created a ramp, and slid the piece off.

Gary does an impression of my father yelling, "I need that ring!"

There's something intimate about the dealer's room. Everyone is trusted as an insider, even me.

A staffer behind me holds a ring on her finger and says, "Six million." I ask if she heard about the diamond they just found in Lesotho, the poverty-stricken enclave kingdom in the middle of South Africa.

The rough diamond was extracted in September 2008 from the Letseng Mine, and weighed in at 478 carats. It is the third mammoth stone to appear in this mine in three years. The other two weigh 603 carats—about as heavy as twenty-one U.S. quarters—and 493 carats. In addition to being very big, this newest diamond is very clear and very white. After it's cut, it could be one of the largest polished diamonds in history.

My father asks Gary if he heard about the diamond. Gary tries to guess which jeweler will buy it. He mentions Laurence Graff, the King of Diamonds, and Lev Leviev. Gary says Graff *has* to buy it, just as Harry Winston would have had to buy it. Graff Diamonds has already purchased the

603 and the 493 caraters. This would make the triumvirate. On the wall near our table hangs a poster of the late Mrs. Harry Winston, printed in 1992 as a memento for Sotheby's auction of her estate. There was a time when Harry Winston was called the King of Diamonds, and a time even earlier when that title belonged to Barney Barnato.

One of the dealers sitting nearby asks my father if I am his new assistant. My father answers that I'm his daughter, that no, I'm not joining the company.

"This is gonna end with me."

When Gary comes back to our table, he has something gigantic and yellow in his hand. "Paul, I want you to take care of your daughter," he says, smiling, and hands me the biggest diamond I've ever even come close to touching, a thirty-seven carater lodged in a ring.

"It's as big as my hand," I say.

Gary tells me it shrinks, but I can't imagine the stone ever seeming ordinary in size. I put it on my finger, just over my wedding band. I can feel the weight of it on my tendons. Then my father takes his turn holding the ring, like a schoolboy admiring pornography in the courtyard. He observes it from a distance and smiles at it, the same smile he wears when he's tipsy or overtired. "Nice lemony color," he says.

Gary gives him another biggie. My father moves it back and forth in his hand. "What a life," he says. "What a life."

Gary believes this stone is a Golconda, from the famous Indian mine. He says he'd stake his marriage on its being the real thing.

Soon, talk turns to the recent Hong Kong trade show and how fruitless it was. One dealer points to Gary and calls him "the litmus test." He's right. Auctions reflect the amount of money people are willing to pay for different categories of diamonds and jewelry.

The men talk about the economy, the banks, the CEOs who are staying wealthy while their corporations fold. The diamond men are anxious. Tomorrow's auction will set the tone for this season's prices. They expect soft numbers.

The next day, I go back to Sotheby's while my father stays at the office. He rarely sets foot in the house on auction day. When he bids, nobody sees or hears him. He sits at his desk on Forty-seventh Street, and a telephone representative bids on his behalf. This way, he can do other work while items he's not interested in are being sold.

The other reason he stays away is auction fever. It's easy to get excited about a piece when other dealers are bidding high, so my father prefers to remain at a safe distance in his office, which is more conducive to even-mindedness. The fever works both ways: if people see him bidding on an item, they might bid higher, as well. Not to mention, he doesn't like having everyone know his merchandise before he's had a chance to resell it. When he bids on the phone, he bids anonymously.

The auction is held in an area of the jewelry floor that has been cordoned off by temporary walls to create a makeshift room. Rows of seats face a podium at the front, behind which hang two large projection screens. Positioned around the room are telephone banks, each seating a handful of operators. Miniature partitions separate their desks from the audience, so that people cannot see the notes they take.

Real auctions are quite similar to those on TV. Gary plays

auctioneer, and although he doesn't say "Going once, going twice," he does bang a gavel on his podium to close each bid. Sometimes he smiles. Sometimes he pretends to be annoyed or happy with the progress of a sale. He knows the names of the people manning the telephones, and when they bid on behalf of a client, he warns them that they are losing, or gives them one last chance with his voice. "I have sixteen thousand dollars here with me, at sixteen. Sixteen thousand. Seventeen thousand at seventeen, but I have eighteen thousand dollars against you now, Geraldine, at eighteen thousand." For absentee bids, written down in his auctioneer's book, he says, "I have twenty-one thousand against you all." A woman behind me bids on an item and wins. Then she puts her sunglasses on and walks out.

Some of the operators are men, but most are female. From my seat, I can hear the telephone girls' S's. Most of the girls are pretty and well dressed. A great many of them are blondes. Some chat with their clients, while their eyes stay focused on the bidding in the room. One of these operators is my father's bidder. I make it my goal to find her.

I have with me a list of lot numbers he will try to bid on, and I follow these items as they come up. The first lot on my list, a round diamond ring, draws bids quickly. In a matter of moments, it is sold for $36,000 to one of the phone bidders. A girl named Katherine calls out the identification number of her phone client, LO207. I wonder if LO207 is my father. I send him a text message. He writes back, "I didn't get it. I stopped at 30,000."

The winner of the bid will have to pay a twenty-five percent buyer's premium on top of the $36,000. Buyer's and seller's premiums are how auction houses make their money. Sotheby's buyer's commissions vary on the lots' sale prices.

Anything under $50,000 commands a twenty-five percent buyer's commission, anything between $50,001 and $1 million takes twenty percent, and anything above $1 million is twelve percent. (These figures are the same at Christie's and Doyle's.) Although this information is made available to the public, it is easy to forget it during an auction, when base prices are called out quickly and sums of money in half a dozen currencies appear on the screens behind the auctioneer. None of these amounts, written or spoken, include the buyer's commission.

The diamond ring isn't the only piece that sells high today. Numerous lots bring in figures that exceed even Sotheby's estimated range. The prices aren't as soft as the dealers thought they would be. In fact, they're hard as ever. Inside the auction house, it is as though there was never an economic collapse: the banks are still standing, the money still flowing.

I turn around to the back of the room and notice a dealer friend of my father's, a woman with a booth in 10 West. In the outer aisle across the room, I see two men in suits conversing. They don't sit next to each other, rather one in front of the other. I think they must be dealers. They have the decisive demeanor that says they are here for business. A few rows behind me, I recognize another Forty-seventh Street man, to whom my father once sold a pair of diamonds. I remember it was an intricate and far-reaching deal. One of the diamonds was located in Brazil and belonged to a man that my father's friend Michael Goldstein knew. The Brazilian diamond's cut, color, clarity, and weight were strikingly similar to one of my father's stones. Michael would broker the stone the Brazilian was selling to my father, and my father would then sell both stones to the buyer for a profit.

The buyer bargained with my father all day over the phone. I listened as they called each other back and forth, my father all wound up.

"I'm already almost sorry that I put these two stones together," he confided to Michael.

Finally, by the afternoon, they'd agreed on a price. It was lower than my father had hoped but probably also higher than the buyer wanted to pay. At the end of the day, we went to the buyer's office to close the deal. On our way out, the dealer said to my father. "Bring me big pearls next time."

"Big pearls?" asked my father. "I'll go up to a lady on the street and—" He made a grabbing motion, as though pilfering pearls off a woman.

At the entrance to the auction area, two men kiss each other on the cheek. Other than the two Forty-seventh Streeters, most of the people in the auction room are unfamiliar to me. The majority of them are likely private customers. Gary told me that in the seventies and eighties, the jewelry auction scene was swarming with dealers. About sixty-five to seventy-five percent were members of the trade. Now it's the other way around, which makes it tougher for the diamond people to procure merchandise. As a general rule, privates outbid dealers. Because they don't need to leave room for resale value, they have more leeway in their spending.

The shift in dealer-to-private ratios reflects a larger change in the auction world. In 1983, Alfred Taubman, a shopping center tycoon, bought out Sotheby's and ordered that it stick its nose back down and welcome common civilians—privates. "People don't want to come in our door, because we make them feel stupid," he said. Under his governance, customer service improved and the auction catalogs became prettier and more useful. The one I was handed at the preview was heavier than a book. Jewelry and diamonds sprawled out on its pages like pinup girls.

The attention that Sotheby's—as well as Christie's—started paying to privates changed the nature of the second-hand jewelry market drastically. Because of auction houses' increased popularity, more and more estate lawyers sold family heirlooms at auction, where they could get higher prices, instead of going straight to diamond and jewelry tradesmen. Rather than being given first dibs on estate merchandise, antiques dealers now had to compete with privates in auctions.

But soon, Sotheby's and Christie's got themselves in trouble. Their buyer's commissions each went up from about ten to fifteen percent. Behind closed doors, in London and New York, the kingdoms were getting friendlier, and several high members of their staffs were accused of price fixing. When the scandal came to light, there were lawsuits. Alfred Taubman went to jail. But Christie's and Sotheby's are still the world's leading houses. And, to this day, Taubman says he's innocent.

About halfway through the auction, I identify my father's representative. A bidder named Eve calls out a winning offer for a diamond ring, and shortly thereafter, he texts to tell me that he bought it. I send back my congratulations. And for the rest of the auction, I can hardly stop looking at Eve. Her hair is short and black. I notice when she is on the phone and when she is not. I notice her eyes glancing around the room and landing on me, and I realize my father must have told her I was there and described me.

Eve smiles while she talks to my father. She looks happy. Perhaps, I think, he is telling her a joke. As in a Greek drama, all the real action at Sotheby's occurs offstage—over the phone. None of the rules of private transactions apply here. If a person chooses, he can remain faceless, nameless, invisible. Bidders register in advance with credit card numbers and driver's licenses. Credit is checked through a bank. No phone calls are made to Forty-seventh Street offices, asking, Is so-and-so okay? Is he good for it? Auction purchases demand no personal relationships, no family connections, no handshakes or small talk, no *Mazal.* Here, price reigns. Auctions are the great leveling fields in the jewelry and diamond world, bringing together buyers from different classes, cities, even countries. Whereas the offices of the diamond district are closed to outsiders, auctions are open to anyone.

A girl in jeans walks into the room with a New York tour book. A blond woman in a miniskirt bids on a diamond necklace with a hundred and sixty-five stones in it and wins. I nod and smile congratulations in her direction. Another bidder looks like a bag lady. Soon, the girl in jeans walks out of the room, beaming, excited. When I adjust my glasses on the bridge of my nose, I realize that if I were a regular, this could be mistaken for an offer. It would not be the first time an auctioneer confused an everyday gesture with a secret signal. It's anyone's game.

From the phone bidding station, one of the reps asks Gary whether he would take $535,000—not a round number.

"Sure. We've come this far, no problem . . ." As though he is handing out freebies.

When the last item of the auction's first session has been

bid on I can see Eve's mouth saying "Thank you" over the phone.

"How'd he do?" Gary asks me excitedly. We are standing in the aisle between the two islands of chairs. The crowd has started to disperse for the mid-auction break. "Did he get anything?"

I answer him quietly, aware of how important discretion is to my father, who won't even come to the sale to stake his claims in person. I'm not even sure I should say hi to the dealer I recognize in the back, for fear of giving away my father's newly acquired merchandise. "I think he got one," I say to Gary.

After the session, I leave Sotheby's and go back to the office. The auction's second half is under way. This time my father gets a call from Carol, another operator. Prices are still unexpectedly hard. "Maybe Jimmy's right," my father says to me—about the world's turning to gold and silver.

When there is a long pause between lots he's bidding on, he hangs up. In the meantime, a client stops by. My father has prepared a tray with parcel papers, diamond boxes, and jewelry. By now these accessories are familiar to me. I know them almost as well as my own belongings. I know that the reason the parcel papers' insides are blue is that diamonds look whiter against a blue backdrop, because the human eye uses blue to neutralize yellow.

While the man is still in his office, my father gets back on the phone with Carol. "Twenty thousand. Bid. Bid. Bid. Uh-huh. I'm out."

Things move quickly today. My father gets on and off the phone. "Hello? Hey, Carol-from-Sotheby's . . . you guys slowed down quite a bit, uh? . . . How the prices doing, still crazy? . . . Well, I thought that was too much money. . . . Yeah, that's what I'd like it at." He laughs. "But we shall see. That might be wishful thinking . . ."

Soon, another man comes up to the office, a broker in a rush. He jokes about my father's many doors, that they're fit for the president. He hands him a diamond, and my father looks at it but doesn't like its craftsmanship.

"Understand!" says the broker. He kisses the mezuzah and runs out, humming.

"That was quick," I say.

My father tells me that brokers need to be quick. Their profit can be as low as two percent. They have to keep selling. "They all have how many kids . . . ten, twelve, fifteen, who knows how? And I don't know how they do it, quite frankly."

"Wooh. Betcha all the Indians are bidding on that," he says to Carol, when a large diamond ring comes up. Then, "Bid forty. Oh no. Oh . . . I'm out."

Off the phone again, he counts how many more items are left. "Almost done," he says to me, and to himself, "Home stretch." Then he's on again and confessing to Carol, "I have to admit I'm not too strong on the last three pieces, so, we'll see. . . . You'll help me steal 'em, uh?" he asks conspiratorially.

During the next interlude, he xeroxes Hebrew psalms from a prayer book. It's September. The Jewish anniversary of Steve's death is approaching, and my father and grandparents will visit his grave with a rabbi to recite prayers. The rabbi gave him a list of psalms, handpicked for Steve. It's been five years now, but my father makes sure to fill his life with reminders of his brother: the picture of Steve on top of the safe, Steve's pottery collection in our kitchen, regular phone calls with Steve's friends in Canada, and, of course, the visits to the cemetery—not only on anniversaries but also before any major trip my grandparents take, as though to bid him farewell again.

As the last lots approach, my father listens to Carol tell him about a restaurant she's going to after the auction. He jokes that if they miss the lot she'll have to buy him a drink. He pouts to her, "I hope no one likes this," and when someone likes it too much for him to compete with, he says, "Aw, okay, I'm out."

The very last lot is coming up. "Gary can breathe easy, had a decent sale," says my father. "Yeah, that's true. . . . So you don't have to go work for Lehman's?" He laughs. "Well, they're not doing it anymore right now. Poor bastards. Four fifty-four? . . . Okay, so everyone's afraid to buy it. Okay. Yeah. Yeah. Yes," he says, excited now, thinking he might have a shot. Then he huffs, frustrated. "That's it, I'm out." He tells Carol to enjoy her drink, and wishes her good-bye.

My father used to purchase ten to twenty pieces at auctions together with a partner. Today, he has bought only two items: a signed Cartier clip with tourmaline and diamonds, and a gold-and-diamond ring, the same one I'd louped during the dealers' preview, the one with the black spot. It wasn't a great sale for him, but it was for jewelry in general. Something amazing was happening at auction houses, not just Sotheby's. Despite the recession, people were buying. Prices were soaring. Even the brooch with the chipped enamel went for $36,250 that day.

"Sorry," I said.

Three months later, Graff bought the 478-carat Light of Letseng, just as Gary had predicted. He paid $18.4 million at an auction in Antwerp. That same month, he bought another major stone, a blue-gray diamond that was once a gift from the king of Spain to his daughter, for $24.3 million, the most money ever spent on a diamond at an auction.

But the smaller dealers were suffering. On October 31, 2008, diamond prices had dropped again, just as they'd feared in Vegas. A round five-carat D-Flawless stone went from $732,500 to $696,000 on the Rapaport list. The market came to a virtual standstill. Dealers stopped buying. Some would become insolvent. Between January and September of 2009, America lost 372 jewelry retailers, out of the 22,623 it started out with. On Forty-seventh Street, *For Rent* signs hung in storefront windows; empty booths haunted exchanges. One of the oldest district restaurants, a small diner on the second floor of an exchange that used to serve boiled carp and canned sardines, closed down.

It seemed that every day I overheard my father talking about someone who owed him a lot of money and couldn't make good on his credit, or a friend of his who was getting held out on. These are burdens dealers have to carry on their own. Tell too many people about a customer's unreliability, and everyone will rush to call in their fees, making it even harder for him to pay you his debt. A wave of uncertainty and suspicion acute even for diamond men overtook the district. Dealers who once trusted each other couldn't; no one was a guaranteed pay.

When I saw Elvis that spring, he sang me a sad song. "Well, I'm tired and so weary, but I must go along, till the Lord comes and calls, calls me away . . ." Most Forty-seventh Street dealers were also tired and weary. The feeble economy had slowed business down to a crawl. While I watched my grandfather and his friends play cards, a man, desperate to sell, asked me if I was a buyer.

That day, Elvis was without his Taking Care of Business necklace. He hadn't worn it in the past eight months. He hadn't been wearing any jewelry at all. The situation had gotten so bad, he told me, "it doesn't give me pleasure to wear fancy things. You would prefer to wear it when you're feeling good and you're doing business, but when things are quiet, you don't have that desire."

Chapter 19

The New Dealers

Most of the diamond people's children are like me. Ben Green's son became a travel agent. David Abraham's son is trying his hand in the entertainment industry, and the setter's daughter goes to medical school.

My father jokes that there are no young dealers in the district; no one wants to join the business anymore. I searched, but all I could find was a handful. Finally, after I'd exhausted most of my references, I started approaching random young men on the street to find the new generation of merchants. This was how I met Joey Cohen one summer day, standing outside the Diamond Dealers Club.

He was dressed in jeans and a striped shirt, sleeves rolled up. His open top button revealed a chain, presumably attached to a Jewish star, and some chest hair. On his head he wore a *kippah,* but he wore it more like an accessory than religious attire; it was made of a fashionable gray velvet and inscribed with his Hebrew name in gold thread. His dark hair was subtly styled, the gel manifest only on the curls at the top of his head. A slight gap divided his front teeth, and a light scruff covered his cheeks and part of his neck. An hour later, we were sitting across from each other at the café of the Diamond Dealers Club, and Joey was telling me how he became a diamond man.

He had toyed briefly with the idea of banking in college. His grades were good enough to get him into grad school, and he conferred with several members of his family about the prospect, but they were split. Joey tried to envision his life if he chose finance: twelve-hour days at the office, stress, "not being able to see anybody, having no life." He hoped to marry young.

On Forty-seventh Street, his father ran a respectable business, earning enough to support his family and still have time for them. So Joey decided that "instead of killing myself going off to grad school, going for my MBA and everything, internships at Goldman Sachs for instance," he would join the family business. While I worked for my father, Joey worked for his, in an office across Fifth Avenue.

After a few months of apprenticing, he and a group of friends enrolled in the GIA to train their eyes in the finer points of the stone they hoped to live off of someday soon.

In their gemology class at the midtown New York campus, they were known as the Rat Pack. There were six of them. Joey was the oldest, the tallest. They still call him Big Brother.

Every morning, the four who came from Great Neck, including Joey, met at the Long Island Rail Road station in time to take the 7:47 train into Manhattan. Two of the boys lived in one neighborhood and two in another; each pair carpooled.

The boys wore slacks when they had work after class, jeans when they didn't. On the train into the city, they rarely talked. They reviewed their notes or studied for tests, or, when they found a seat, they sometimes slept.

When they arrived in Manhattan, they walked to campus and spent their day, from nine to five, becoming diamond experts. They sharpened their eyes to the subtleties of color and learned to identify the different species of blemishes and inclusions: feathers, fractures, clouds, and crystals.

"We were at the top of the class, all of us," said Joey. "'Cause we all came in with experience. Our fathers were all in the business."

This was how the Rat Pack spent the summer of 2008. While the rest of their Long Island friends hung loose at bars and parties, Joey and his diamond gang studied hard. "Friday nights, just come back and crash."

"Yeah?" I asked.

"Crash," he repeated, shaking his head. When he said it, the word itself sounded like a plane hurtling into water.

Joey was used to putting in long hours. He had taken summer classes throughout college so that he could finish his studies in three years and join the workforce. That summer at the GIA, he didn't just learn about the physical properties of diamonds. He became versed in the anxieties of being a diamond man. "I remember one time, one of the stones I was grading, it jumped from my table, from my tweezers—'cause I probably squeezed it a little bit too hard, whatever—jumped off my table, and I lost it for like half an hour, and I started sweating. My heart was throbbing." In some ways, losing a stone at the GIA is worse than losing a diamond you own. If you are a dealer on Forty-seventh Street, chances are you acquire your diamonds at marked-down prices. But a student who loses a diamond the GIA lent him to practice on must reimburse the lab at full Rapaport cost.

"I literally pulled out the carpet in the classroom until I found it in the corner, in a crack somewhere. So yeah, that was a very, very scary moment, and it happened a lot of times. You're used to it. It always happens. Diamonds fly everywhere. Hold 'em with tweezers, you lose them. They're very small."

After the GIA course, the members of the Rat Pack all went their separate ways. Only three of them, including Joey, are based on Forty-seventh Street. Of the six, one deals in jewelry, another in colored gemstones. A third specializes in melees—diamonds weighing less than .20 carats. Joey himself buys and sells larger stones, starting at .80 carats. The largest diamond he and his father acquired was 34 carats.

Joey is still close with the others. They come to him for advice, both diamond and otherwise. Once when the dealer life had been treating one of them harshly, he called Joey, who left his father's office and went over to his friend's father's office right away. They strolled around the district together.

"I went, 'Listen, you're going through growing pains right now, that's all it is. It happens to everyone. Just that it's good to have someone there for you, talk to you, get you through it, and that's it. Someone you trust.'"

Joey knew what it meant to be a greenhorn. He remembered his first few months on the street, before he'd been to the GIA, before he could navigate those minute differences between an H- and an I-colored diamond. It took him a long time to gain respect among the dealers. Some saw Joey not as a salesman but as the son of a salesman. Some days he wondered if he would last in the business.

"It was very depressing. I wasn't selling anything. I was coming here. I felt like I was wasting my time."

By the end of his first year on the job, Joey had discovered something that only he could offer his father's company: the Internet. By joining RapNet, Martin Rapaport's online diamond-trading network, he increased the company inventory fourfold.

At first, his father resisted dealing diamonds online. Computers were not his friend. "Couldn't even find the on-button on the computer."

But the real issue was that Joey's father "is more a face-to-face business of a guy," someone who likes to physically examine a stone before he buys it, to approve it in the flesh, so to speak.

"I came in, and I'm like, 'Dad, let's put your stones on the Internet. Let's see what happens. You have nothing to lose. You pay a *little* bit of a membership fee. It's *nothing.* You sell one stone, you make it right back.'"

The Internet is what Joey considers the main difference between himself and merchants of his father's generation. "By far the Internet. These guys had no idea what was going on."

Joey's faithful position behind the computer screen made him the subject of gossip among some of the old-timers. One of his grandfather's friends confessed to being approached by others. People asked him, "What is this kid doing here? He's just sitting down on his computer doing nothing."

The man had loyally defended Joey. "What do you care what he's doing or not? He's here. He's making his money. He's doing his business."

"They don't know what's going on behind my screen," Joey said to me. "Only I do, my father does, and the people that know me. Now they respect me. They come up to me. 'Do this for me.' 'Can you check this stone out for me?'

This and that. Now, once they're familiar, they start trusting. Obviously there's a learning curve. It's new blood and everything. I'm the rookie. Everyone distrusts the rookie. There's always a level of trust you have to build up."

For Joey, the Internet is also a bargaining tool. If a dealer tries to sell him a stone for too much, he will point to a comparable diamond on the Web that costs ten percent less than the seller is asking. "Obviously here it's gonna be easier, 'cause you're dealing face-to-face with somebody, but I have the Internet in the palm of my hands." Joey brought his hands together. "I can go get a stone from L.A., ask him to ship it for me. Yeah, it's gonna cost what, fifty to two hundred dollars? For ten percent cheaper? Obviously I'll go that way."

A Belgian had seen his diamond on the Web, had called to say he was wiring the money at once, and could Joey mail the stone to Europe? "That rarely happens that someone buys a stone without even looking at it, without even looking at the certificate. 'I wanna buy this stone. I'm wiring you the money.' He didn't even bargain with me. 'Here. Here's the money. Give me the stone. Send me the stone ASAP.' He wired me the money that day, sent out the stone the next day. Done. That would never happen with someone face-to-face in my father's older generation."

The Internet gave Joey his first sale, something he told me he would never forget.

When I asked him what it felt like, that first deal, he said, "Aww," breathily, as though conjuring up a goddess. "It was fantastic. It's like wow." He had survived those first few barren months in the district—no deals, no phone calls even. But now he was an earner. "Once you see the money, it's worth it."

He celebrated by taking his family out to dinner at a kosher restaurant in Great Neck. "It's such a great feeling just to know that it's not like my dad's throwing money at me, you know what I mean? Like I'm *making* my own. I'm *expanding* his business. I'm taking his business and I'm making it more efficient."

There is something else that's won Joey esteem among the dealers: his eyes. Before he took his GIA course, the ins and outs of diamond quality were still a mystery to him. Color was the hardest of the attributes. "Okay," he laid it out to me. "So when you look at a stone, you could tell what's wrong with it. You look inside, trained eye, you'll be able to see, 'Oh, there's a black thing here, there's a white feather, there's a white crack over here.' Fine. Color." He stopped with gravitas, as though recalling an infamous adversary. "Color," he said again, shaking his head, "was very difficult to tell, because you had nothing to base it with. There's nothing to compare color with. Another thing also, still I'm having trouble with the pricing of stones. That's a very big, big, big problem for me right now. See, I'm very inexperienced in my third year working now. Prices are fluctuating here and there."

In a faltering economy, price is a challenge for everyone. But Joey has managed to sharpen his eye for color so dramatically that his father regularly consults with him on diamond grading. "Even other people will come up, like, 'Joey, what do you think this stone is? Can you grade it for me?' I'm like, 'Yeah, sure. No problem.'" His father has told him that he is probably one of the best diamond graders in the neighborhood.

When you are twenty-four and other dealers seek your advice on the color of a diamond, it hints at an aptitude that surpasses a summer course at the GIA, a natural talent that exceeds the average dealer's knack. I asked Joey if he could trace his talent back to anything. Tentatively, he mentioned video games. He said, not without thanking God, that he has good eyes. "I used to play hours and hours, every night, video games. Every day. Up to college even. Three hours a night sometimes with my friends." He thinks the games probably trained his eyes in spotting details. You're always on the lookout for tiny keys to another world and hidden enemies. Not unlike inclusions. The new generation of diamond dealers emerges fresh from the world of Halo and iPods.

Joey's first memory of the business is accompanying his father to the London Diamond Bourse in Hatton Garden, the British equivalent of the New York Diamond Dealers Club and Forty-seventh Street. His family lived in London until Joey was nine, while his father pursued a career as a cleaver and, eventually, a dealer.

"I remember those huge tables everywhere, phones everywhere. It was very overwhelming for me. People running around, this and that."

According to Joey, "Kids always think they're gonna end up doing something like their father. Obviously, all kids have dreams to be something else." Joey's was football. But he knew that most likely he would join the family business.

I considered it a profound realization for a young boy to come to; to understand that he might fantasize about pigskin and touchdowns, but when it came time to earn a living, he would suit up and join his dad. Joey, I thought, became a dealer long before he sold his first diamond. He became a dealer when, at the age of twelve or thirteen, he chose a merchant's life.

Age seven is probably the point where I fell behind those boys and girls who would later grow into the successors of their fathers' diamond businesses. I was not one of those kids whose fathers conveyed them into their offices on snow days or Take Your Daughter to Work days to begin fine-tuning their eyes to diamond color. It has always been clear to me that I was not destined to be a diamond dealer. But out of morbid curiosity, I asked my father one day if he thought I'd make a good one.

He let out a deep breath. He said he hadn't thought about it. I could tell he was worried about offending me. He has always thought of me as sensitive, breakable.

"Probably yes and no," he said. "No, because you're not aggressive enough. And the ones that make it are more aggressive. And yes, because you have a really good way with people. And you're smart. So from the personality point of view, which is very, very important in our business, yes. And you're certainly smart enough to do it," he repeated, "and smart enough to learn it." There was only the matter of aggressiveness.

But Joey wasn't aggressive. In the Club, he said dotingly, "It's so much fun, just to see the beauty of a diamond, to see the reflections, this and that, how you could tell *so* many different types of diamonds, different types of cuts and everything. It's just"—he paused—"a diamond is beautiful. I love to look at it." My heart melted. I hadn't expected a romantic.

But he was definitely also a merchant. Shortly before, a man had walked by our table at the café and handed Joey a parcel paper, which he slipped into the front pocket of his shirt. His favorite aspect of dealing is the fast pace of transaction. Sometimes he and his father will sell a stone they've bought just an hour before. "Someone comes to my desk.

'Do you have this stone?' 'Yes, I do. Here, sold.' Done. Matter of minutes. Just like that. Profit."

On an average morning, Joey drives his father to the train station so that his father can board the 9:22 to Manhattan. After seeing him off, Joey parks the car and catches the next train, ten minutes later. This detail bothered me for a while, but when I asked him after a few months why his father couldn't just wait for him and take the same train, he answered simply, "Because the 9:32 is a local train while the 9:22 is express," as though it were obvious. I had also come to understand that there wasn't always time for pleasantries in a diamond office. When I first started working for my father, he had warned me, "Sometimes I'll have to raise my voice if I'm in a rush. I hope you'll understand and not be hurt."

The first thing Joey does when he comes into the office is boot up his laptop. He uploads descriptions of the company diamonds onto RapNet, walks some stones over to the GIA lab on Fifth Avenue for grading, or stops by the offices of potential customers who want to have a closer look at his merchandise.

As Joey spoke about his day, the intermittent din of a nearby chess game between Club members sounded from the back of the café. "Aren't you embarrassed? Aren't you ashamed of yourself?" one man demanded of another.

Today, the lounges, where card and board games are played, are the most animated part of the floor. There was once a time, in the seventies and eighties, when the Diamond Dealers Club was hectic with brokers at nine thirty in the morning, when people bought jewels like candy, and the diamond men and women rushed to their tables to open up shop. "Now," one dealer told me, "it's like a ghost town." People mosey in late in the morning. Sometimes they don't come at all. Outside, hawkers roam the street, calling desperately for customers.

When I met Joey, the district was undergoing one of its worst dry spells in history. Between the spring of 2007 and the summer of 2008, twelve tenants left one of Forty-seventh Street's office buildings because the rent became too steep for them. The building's owners planned to welcome non–diamond enterprises. Another district condominium had changed its rules to do the same. If too many outside businesses replace jewelry companies that can't afford to stay, the district will no longer be the diamond mart that it's been since the 1940s.

"This is the only district left in the city," said Ken Kahn, the landlord of 10 West. "The flower district is gone." He counted the disappearing districts on his fingers, pinky first. "The garment district is gone. The meat district is gone. The toy industry is not here anymore. The gift building is now a condominium. Forty-seventh Street is the only district." Yet, as the diamond industry modernizes, there is a tension between rapid innovation and those archaic traditions that have persisted. This tension is epitomized by what has come to be known as the International Gem Tower.

In the early years of the millennium, a new character entered the cast of Forty-seventh Street. He envisioned a modern diamond district, one closer to the international trading complexes of Tel Aviv and Dubai.

His name is Gary Barnett, though in Antwerp, where he started out as a diamond dealer, people called him by his Jewish name, Gershon. Barnett made the crossover to real estate in the nineties, when he moved to New York. By 2006, Extell Development, the company he chairs, had seven developments going. He dreamed of building a diamond tower on Forty-seventh Street that would house all parts of the trade: the GIA, the Diamond Dealers Club, banks, a secure garage, and, of course, the diamond men and diamond women. Or maybe it was Gershon the diamond dealer who dreamed this.

Pretty soon, he'd gotten a number of Forty-seventh Street landlords up in arms by approaching the city government for subsidies. They joined together and hired attorneys, a lobbyist, a publicity firm. A few years later, Barnett got a package of $49.6 million in tax breaks from New York City and State anyway. New York State will also provide a fund for mortgage financing of up to $100 million for people to buy condos in the tower.

Many dealers are for the tower, hoping it will revitalize the neighborhood. But for me, it is bittersweet. I'm not sure I prefer a district that actually looks like a place where millions of dollars in diamonds course through the streets, instead of a medieval market. Maybe more privates will roam the vicinity, pressing their noses up to improved window displays without ever hearing the words "We buy diamonds" mumbled into their ears like a secret, without ever having to turn down an offer to be ushered into a walkup store they wouldn't otherwise know existed. But what about the street deals and the hawkers, the men with signs strapped over their torsos and the storefront tickers announcing *We pay cash*? I worry about who will be exiled from Forty-seventh Street when they cannot afford a place in the newer, nicer district, about all that would be lost if they left this diamond world.

Construction has already started on the south side of Forty-seventh Street, closer to Sixth Avenue, and the GIA has decided to relocate its laboratories to the tower when it opens. But the Diamond Dealers Club isn't moving. It renewed its lease with 580 Fifth Avenue, the building where so many of my father's friends have their offices, the largest diamond building in the world for now. When I stopped by the Club in the summer of 2009, suspension, expulsion, and resignation notices overwhelmed its bulletin boards. I heard my grandfather and some other Club members gossiping about a man on the street who had recently absconded with a million dollars' worth of goods.

Still, the Club has about two thousand members. Its trading floor is an important place for all kinds of diamond merchants, including Joey, who has made several impromptu trades just because he and another dealer were on the tenth floor of 580 at the same time. Every once in a while he arranges to meet someone at the Club to view or show a stone they've previously discussed on the phone or Internet.

It's good to be seen there, especially if you are a young dealer, someone who is trying to make a name for himself, because despite the popularity of Internet sales, the DDC is still the core of Forty-seventh Street and an office for many New York traders. Some members even print its number on their business cards.

When he's at the Club, Joey will check the bulletin boards to see if any out-of-town buyers are visiting and what sorts of goods they're shopping for. "Then I match it up with my inventory, whatever I have, bring it up, and show him, and I say, 'I'm the next up.'" Usually, there is a line when an out-of-towner is there. So Joey inquires among the gathering of dealers and brokers, "Who am I after?" and then takes a seat, making sure not to get too close to the man currently flashing the buyer his goods. He waits. "Privacy is a big problem here, very big problem here, but you learn to deal with it," he confessed to me. Then he qualified this. "But no, everyone's very respectful of each other. Everyone knows their place. They know what to say, what not to say at which times. If you're doing business, someone's not gonna come up and look over your shoulder and see what's going on. They're gonna respect."

It was when I asked Joey what he thought of the Rapaport list that I began to really understand his generation of dealers.

"In terms of what?" he asked innocently.

I started to say that some of the old-timers are against it, that they feel it tells them what to do, but Joey interrupted.

"It dictates prices," he said casually. "What else are you gonna go by?"

The list, for him, is so absolute that what he thinks about it is beside the point. It is like asking him what he thinks about traffic lights or taxes. He has never seen life without the list. He wasn't there when Rapaport was attacked for having the gall to assign an objective number to something as complicated as a diamond.

For that matter, he wasn't around when widows roamed the Club, and when I asked him what it was like working in such a male world, he said, "Never thought of that." He told me that there were about five girls in his fourteen-person GIA class, that he doesn't see gender.

In his few years of work on the street, Joey has already witnessed the times turn on diamond dealers, practically halting trade. He's also seen the first signs of recuperation, people slowly emerging from dormancy, wanting to deal again. In the beginning of 2010 only half as many American jewelry wholesalers were lost as had been in the first six months of 2009.

Sometimes he still worries about a total crash in prices, but he is young enough to weather the storm. He is on the winning side of economic natural selection. It is the old guard, and the older guard, whose businesses may not survive. But for Joey and his friends, this year will surely not be their last. They are not getting kicked out of their exchange booths; they don't have exchange booths to get kicked out of yet. They are fresh enough to consider this debilitating crisis something as harmless as growing pains.

Joey didn't survive a war to get to this street, did not sew any stones into the lining of his clothes before crossing an ocean. The only thing he's ever had to fight is a slow economy. But he is just like Opa Yankel when Opa Yankel alighted on the street of diamonds, and my father when he set out into the German countryside to hunt for customers. He is just like those first diggers in South Africa to push past the wet boundaries of the sea and deep into the land in search of luster; for him, the world is fresh, open, and glistening with possibility.

One day in the distant future, Joey will take over his father's company, running it with the skills he's learning now. He has already decided that he will buy the girl he marries a cushion-cut diamond, because he prefers them over the more common "run of the mill" round shapes. When he told me that they were his favorite, I couldn't resist mentioning that my father deals in cushions.

The diamond may come from anywhere in the world, and Joey will find it with ease, perhaps without even leaving his computer. This is what the future of the industry looks like, and it is neither scary nor surprising to Joey and his peers. They have all learned to say *Mazal* and work their way through a cachet, but they also have tools that the old-timers could not have conjured up in their most imaginative moments.

The old-timers still dominate the diamond district for now. Nishan Vartanian, an antiques dealer in his thirties, sometimes feels as if he is a member of his father's generation instead of his own. Most of his colleagues are his father's age, some his grandfather's age. The people Neeraj Rawat does business with on Forty-seventh Street are also mostly "older people."

But while the median age of Forty-seventh Street may not be changing as rapidly as in other businesses, its face is. Joey represents the continuity of what the district used to be, but I met a lot of dealers who probably wouldn't have been part of the community a few decades ago. About five hundred of Forty-seventh Street's offices are Indian, one of them the Rawats'.

In 2009, Efraim Reiss, the Bobov Hasid, introduced me to a colleague of his, a Druze Muslim Arab from Lebanon. Samer had a nice big office in 580, where he welcomed me.

"People ask me, because I'm always between the Middle East and New York, so people ask me, 'How do you manage to crack into it? How do you manage to do business with them? Actually, is this so difficult? Or do they hate you because you're an Arab?'" But Samer tells people that the dealers of Forty-seventh Street are his family. "And it's not what the media says on CNN. It's a different world. People, they don't judge you based on your religion. People judge you based on your behavior and who you are and if you are trustworthy or not. They don't care—at least the way I see it here from people that I know—they don't care what's your religion or what you do. They even called me to bar mitzvahs, to events. I shared with them their happiness and I shared their bad times, too, you know. We've been to *minchas* together, I put *kippah* on my head."

In his safe, Samer keeps the mezuzah that hung on the doorpost of his first Forty-seventh Street office. When the old Jewish man with whom he had shared his workplace asked why he was taking it, Samer asked the man if he thought the mezuzah distinguished between Jew and non-Jew. "This mezuzah," he said, "is good luck for everyone, so I'm taking it before even I take my safe."

Like the mezuzah, the Jews that Samer has encountered on Forty-seventh Street don't really distinguish between Jew and non-Jew either. This surprised me. It surprised Samer, too. But about two or three years after he joined the Club, he and another dealer got into a disagreement and had to take their dispute to arbitration. Samer was sure his rival would win the case. The man was "close to the people over there and I was new in the Club, and I was worried because they may find me guilty, because you're all Jewish together and all that," but after they called the two litigants back in, the arbitrators explained that Samer was to be compensated. They had ruled in his favor.

The traditions of the street, of the diamond business, have become Samer's. Like all the others, he says *Mazal.* "*Mazal,* it's an international word."

Forty-seventh Street has become international, too, and it stands on the cusp of transformation. I'm not sure how long before the borders of its world break and the last strands of antiquity give way. If I joined the business, I'd inherit a very different world from the one my father did. I would call laser workers, not cleavers. I would trade gems over the Internet. I would walk through the halls of Barnett's new building, and maybe, as I crossed Fifth Avenue from the east side to the west, I would no longer hear the calls of the hawkers. By the time I came into my own, Mr. Green would be retired, and perhaps carrying jewels in chest packs would be regarded as dangerous and obsolete. In a way, the business is just like its gems—men coming and leaving, their fingerprints disappearing.

When I left Samer's office, there were two men with skullcaps sitting on his fine leather chairs, waiting for him. No matter what changes, the business will always be this: a few men, hoping to turn some glitter into a living.

Epilogue

My engagement ring came from the exchange in 10 West. The diamond was my boyfriend's mother's. It had rested in a simple gold setting when he knelt down in our Maryland apartment to propose to me, and I was getting it reset. My father knew a man who made antique-style engagement rings, so one day in June 2006, I met him in the district, and we walked to a small booth in the leftmost aisle of the exchange. The jeweler would fit my future mother-in-law's diamond into an antique-looking ring. It was my first time as a customer on Forty-seventh Street. I bent down to look at the rows of rings inside the showcase laid out in assorted little boxes, some with bronze clasps, others with painted borders. The jeweler took them out for me, and I studied their details. They were tiny works of art. When I had narrowed the collection to a few models I liked best, I jotted down their code numbers on a slip of paper that I left in my father's office.

The next time we went back, the jeweler asked me to choose a ring size. My father was convinced I needed one that hugged my finger tightly. To sway me, he drew up vivid scenarios involving soap bubbles and sweat. He impressed upon me how easily this ring could slip off, lost forever. In his warnings, I recognized some remnant of the old guardian's vigilance that had guided him his whole life as a father—when he checked my body for ticks, when he kept me from the Israeli Day Parade, when he refused to give me the combination to his safe the one time I'd asked, so that I would not find myself in a position of compromise should worst come to worst, should a robber break in and demand of me information that a less mindful father may have entrusted to the ownership of his young child. It was harder to say no to him since Steve died. I saw the concern, the worst-case scenarios fluttering through his mind, and I gave in and took the smaller size. But I chose the design: an art deco platinum with three small diamond-encrusted steps on either side, leading up to the main diamond. Clean and simple and lovely.

That was the same exchange where just a few weeks later I would pick up my father's diamonds from Ginsburg. I think of that day as my initiation into the trade, as close as I would ever come to being a dealer. It was the first time my father would ask me to take care of his diamonds for him, and putting on the pack, I would come to understand many things about his life that had been a secret to me for so long. If you asked me before I learned that light can leak out of a stone if you cut it a millimeter too deep; before I'd started to call diamonds things like brilliant or quiet; before I'd heard my father and his friends talk about them like art critics; and before I'd fallen in love with the old world, I'd tell you that diamonds were a business like any other. But ever since I carried those stones, I've watched the diamond men and women closely, knowing that each one could be wearing an entire scintillating wardrobe beneath their clothing that no one could see. I would understand that, in a way, being my father's daughter and coming from a diamond family, it was as if I had been wearing the chest pack my whole life, that once you've worn the jewels, you act as though you're wearing them always. I would realize that the reason there are so many flashing signs in a district full of gems is that on this block, the glint of diamonds in itself is not enough to attract attention—there are just too many of them—and that, at the same time, after you've been looking at diamonds for so long, in all their limpid intensity, everything else feels hopelessly unvarnished. But that day, standing on the outside of the booth next to my father, trying on my engagement ring for size and beauty, all of this was still hidden from me.

Notes

Chapter 1: Our Brightest Blazes

Barak Richman's written analysis of the etiquette, traditions, and legal workings of Forty-seventh Street (among them, "How Community Institutions Create Economic Advantage," *Law & Social Inquiry,* Spring 2006) provides insight into the complex codes of business and unique communal ordering that shape the district. He uses the term "diamond-studded paupers" in his discussion of the diamond community. Both Richman and Renée Rose Shield (*Diamond Stories* [Cornell University Press, 2002]) write about the importance of family and reputation in diamond dealing, and Shield notes the presence of oral contracts in the business. *The Diamond People,* by Murray Schumach (W. W. Norton & Company, 1981), offers a look inside the district of the seventies and eighties and a history of the business.

Chapter 2: Glitter Becomes a Currency

Stefan Kanfer's *The Last Empire* (Farrar, Straus & Giroux, 1995) is a thorough guide through the birth and history of De Beers, the characters and lives of Cecil Rhodes and Barney Barnato, and the South African diamond story. The quotes "Gentlemen, this is the rock upon which the future of South Africa will be built" and "it is our duty to take it" (from Rhodes's will) appear on pages 27 and 66. The line "and Barney too!" appears on page 38 of *Diamond,* by Matthew Hart (Fourth Estate, 2003). In his book *The Heartless Stone* (Picador, 2006), Tom Zoellner includes the detail about Barney Barnato's unique soliloquizing style. My study of early diamond trading and its Indian and South African histories drew upon Zoellner's book; Shield's book; *Blood Diamonds,* by Greg Campbell (Westview Press, 2002); *The Book of Diamonds,* by Joan Y. Dickinson (Dover Publications, 2001); Hart's book; the American Museum of Natural History's "The Nature of Diamonds Exhibition Notes" website (www.amnh.org/exhibitions/diamonds/); and *The Nature of Diamonds,* edited by George E. Harlow (Cambridge University Press, 1998). Brazil's historical relationship with the diamond is covered in the latter three sources. Hart, Zoellner, and Kanfer supply accounts of the fight for De Beers. "The Nature of Diamonds," Kanfer, and several contemporaneous newspaper articles (e.g., "Barney Barnato's Suicide," *New York Times,* June 16, 1897, and "Follows Barnato into the

Sea," *Chicago Daily Tribune,* June 16, 1897) document Barnato's declining mental condition and subsequent death. Kanfer and Zoellner contributed to my understanding of what life was like for miners at the De Beers mines after consolidation, as well as the Matabeleland affair. Campbell's book states that De Beers controlled 90 percent of rough diamonds when Rhodes managed the company. Encarta's explication of the Natives Land Act ("Natives Land Act," Microsoft Encarta Online Encyclopedia, 2009) lends insight into some of the particulars of South African racial legislation. Both Kanfer and Edward Jay Epstein's book *The Diamond Invention* (originally published by Simon & Schuster as *The Rise and Fall of Diamonds,* 1982) aided my understanding of apartheid South Africa. Epstein's article "Have You Ever Tried to Sell a Diamond?" (*The Atlantic Monthly,* February 1982) examines the De Beers/Ayer advertising campaign and how De Beers essentially "invented" the modern diamond. It provides many of the numerical figures associated with the campaign's success. Kanfer's book includes some details on Ayer and De Beers, as well. The De Beers advertisements that I describe in this chapter were retrieved from the N. W. Ayer Advertising Records at the National Museum of American History Archives Center of the Smithsonian Institution (Collection 59, Box 40, Folders 4, 5, 6, 7, 8; Box 41, Folder 10). The advertisement that features the quote beginning "Endless they seem to young people caught up in love's first fine awareness" comes from Collection 59, Series 4, Box 4, Folder 4. The quote beginning "Industrial diamonds—a key priority for high-speed war production" appears on several of these ads. I also consulted boxes 1, 2, 4–9, 40–46, 88–90, and 148. Hart and Campbell write about the challenge Canada posed to De Beers in the past. Several sources, including Epstein's *The Diamond Invention* and Kanfer's *The Last Empire,* discuss Harry Oppenheimer's international business arrangement with the leaders of African countries and with the Soviets. Epstein and Zoellner both observe that De Beers made diamonds seem rare despite their geological frequency. Hart describes De Beers' sorting and sightholder processes. The De Beers Group website contains a wealth of information on the company. Campbell, Hart, Zoellner, and Shield write about the transformation of the diamond pipeline. In particular, Zoellner adds an interesting recounting of the Argyle Mine confrontation and De Beers' foray into retail. "The Diamond Game, Shedding Its Mystery" (Lauren Weber, *New York Times,* April 8, 2001) elaborates on some of the particulars of the De Beers transformation, including the revenue-advertising ratio sightholders were encouraged to adhere to. Hart's book educated me on the competition that BHP posed. The quote "master craftsmen," in relation to Forevermark diamonds, derives from the Forevermark website (www.forevermark.com).

Chapter 3: The Deal

Dickinson's book contains a lot of detailed information about the world of diamond cutting—for example, that each factory uses a unique mixture for its cutting wheels. The Gemological Institute of America's website (www.gia.edu), Dickinson's book, and *How to Buy a Diamond,* by Fred Cuellar (Sourcebooks Casablanca, 2008), contain explanations of diamonds' anatomies and characteristics, as well as an abundance of gemological terminology. I learned that Tepper Galleries was the oldest auction house in New York from its website (www.teppergalleries.com). In addition to the informal schooling I received on this topic from spending time in the diamond district, I was educated by Shield's account of bargaining at the DDC and of the Club as a theater with spectators. It is from her book that I learned of the broker who recited Kaddish in response to an offer. Matthew Hart defines the term "gletz" in his article "How to Steal a Diamond" (*The Atlantic Monthly,* March 1999).

Chapter 4: Carbon

Professor Andrew Campbell, at the University of Chicago, spent generous amounts of time educating me on the science of diamond formation. My understanding of this process and of the gemstone's properties was enriched by the works of Zoellner, Hart, Shield, and Robert M. Hazen (author of *The Diamond Makers* [Cambridge University Press, 1999]), as well as "The Nature of Diamonds" website and Rapaport's Fair Trade website. Dr. Yan, of the Carnegie Institution, and Shield's book, were particularly instructive on the topic of diamond coloration. Partnership Africa Canada, www.diamonds.net, and the Diamond Development Initiative website (www.ddiglobal.org) are all valuable sources on artisanal diamond mining. *The Diamond Road,* a movie produced by Kensington Communications, brings to life the plight of diamond miners. Other useful articles include Rob Bates's "Does Diamond Mining Hurt the Environment?" in JCK's *Jewelers Circular Keystone* (May 13, 2009), Dorothee Gizenga's "The Kimberley Process Is Not Enough" in *Modern Jeweler* (October 2008), and Alexa Brazilian's "Hope Diamonds" in *Elle* (December 2008). The postponement of aggregation's transferal to Botswana is covered in "DTC Botswana Shuts Down, Postpones Aggregation to Gaborone" (*Sunday Standard Reporter,* February 7, 2009). The Gemological Institute of America was a great resource on the process a diamond goes through at the laboratory. "The Nature of Diamonds" includes scientific information on alluvial diamonds and on mining. The De Beers Group website offers a comprehensive description of the company's mining, sorting, and sales processes. The quote "The diamond industry's center of gravity is shifting" by Nicky Oppenheimer comes from "Botswana Country Profile" (BBC News, July 27, 2010, http://news.bbc.co.uk). The Diamond Manufacturers and Importers Association is a resource on cutting trends.

Chapter 5: Hashem's Diamonds

Shield writes extensively about Hasidim in the diamond district. In his article "A Sparkling Life" (*The Jewish Journal,* August 14, 2003), David Geffner links the phrase *Mazal und brucha* to Maimonides and his gem-dealing brother. Schumach's, Shield's, and Zoellner's works discuss the parallels between the diamond business and Jewish tradition. Schumach, in particular, discusses the role of religion in the evolution of the diamond industry. John Bainbridge and Russell Maloney's article "Diamonds" (*New Yorker,* November 29, 1941) notes that the Low Countries housed the most burgeoning cutting hubs. The story of the birth of the New York Diamond Dealers Club—and the tale of the lost diamond associated with its beginning—is related in Albert Lubin's *Diamond Dealers Club* (Diamond Dealers Club, Inc., 1982). The New York Diamond Dealers Club website provides additional information on the bourse. I learned most of what I know about the Jaroslawicz murder from Schumach's writings. Several newspaper articles, including "Following Up" (Joseph P. Fried, *New York Times,* September 14, 2003), discuss the event, as well. My timeline of the Jewish diamond experience draws upon information provided by the works of Zoellner, Shield, Kanfer, Hart, Schumach, as well as Professor Dan Michman's *Daf Shvui No. 172* (Bar-Ilan University, *From Turquoise, Sapphire and Diamond to Mazal U'brachah* [Rühle-Diebener-Verlag, 1979]), the Encyclopedia of Genocide and Crimes Against Humanity, and others. Shield and Zoellner write about the presence of survivors in the business. Richard Gibson Hubler's article "Diamonds Come to America" (*Harper's Magazine,* May 1941) offers a contemporaneous account of the immigration of diamond men—and their gems—to New York City. The quote beginning "This year" appears on page 583 of that issue. In addition to my own experience aboard the bus to Monsey, I was educated by Shalom Auslander's enjoyable account in "The Third Look: On Rereading Leonard Michaels' *I Would Have Saved Them If I Could*" (www.nextbook.org, July 29, 2005).

Chapter 6: The List War

Martin Rapaport was extremely generous in relaying his experiences in the diamond industry. His website and articles are also sources of information. Kanfer, Zoellner, and Shield write about the unique circumstances of diamonds in the late seventies and early eighties. The writings of Shield, Hart, Zoellner, Russell Schor (*Connections* [International Diamond Publications, 1993]), and various newspaper articles all contributed to my understanding of the Rapaport list's history. Shield, in particular, discusses how dealers blamed Rapaport for decreased profit margins, a fact that was reinforced during my conversations with industry members. Shield and Zoellner also write about the effects of the GIA and the Internet on the industry. The quote "I became kind of like a scapegoat" was derived from Rob Bates's article "DDC Celebrates Diamond Jubilee: An Oral History" in the *Jewelers Circular Keystone* (July 2006). The quote "Forty-seventh Street has to recognize that it's in America" comes from Sandra Salmans's article "A Diamond Maverick's War with the Club on 47th Street" (*New York Times,* November 13, 1984). Shield discusses the increased prevalence of lawyers at the Diamond Dealers Club and the Club's apprehensiveness regarding the removal of members subsequent to the Rapaport affair. The Rapaport office provided information on diamond prices and trends. Schor mentions the resolution that passed at the World Federation of Diamond Bourses forbidding members from putting out price lists.

Chapter 7: The Older World

Zoellner and others discuss the birth of the famed slogan "A Diamond Is Forever" in 1948.

Chapter 8: A Dealer's Life

My father was a valuable guide on the process of cutting a diamond and his own cutting education. The De Beers Group website contains a host of information about rough diamonds, and the Gemological Institute of America and Joan Dickinson's book are resources on diamond properties, as well. My uncle's written descriptions of his experiences enriched my appreciation of the timepiece market in the nineties. Trade websites such as www.horologist.com contain useful information about a watch's anatomy and the terminology that goes along with it.

Chapter 9: The Diamond Women

For further reading on the status of women in the Diamond Dealers Club during the 1970s, consult "Women: New Facet of the Diamond Trade" (Jane Friedman, *New York Times,* February 18, 1979), "Diamonds Are a Girl's Best Friend" (Susan Linee, *Washington Post,* April 6, 1979), "The City's Most Exclusive Club" (Murray Schumach, *New York Times,* May 6, 1979), and Murray Schumach's *The Diamond People.* Both Shield and Richman write about the unique function that the walls of the Diamond Dealers Club serve.

Chapter 10: The Last Diamond Cleaver

The article "World Traders Move to Curb Diamonds from War Zones" (Reuters, *New York Times,* July 19, 2000) includes useful diamond-cutting facts. The Diamond Manufacturers and Importers Association supplied me with information on the current state and geography of manufacturing. The majority of my knowledge regarding the way in which this craft and business has transformed over the years comes from experts I observed directly and spoke with. The story of Joseph Asscher and the Cullinan diamond is recounted in Matthew Hart's book. Dickinson, "The Nature of Diamonds Exhibition Notes" website, and the Gemological Institute of America are fine sources on the history of diamond cutting. Donna Jean MacKinnon's "Diamonds: Then and Now" (*Toronto Star,* November 4, 1999) is an interesting article on the topic. Harlow writes extensively on the history and role of the diamond in the context of royalty, religion, and mythology, and the transformation of the diamond's significance throughout history. The American Museum of Natural History website, Hart, Zoellner, Shield, and Meghan O'Rourke's article "Diamonds Are a Girl's Worst Friend" (www.slate.com, June 11, 2007) add to this subject.

Chapter 11: The Diamond Detectives

Additional information on the Tiffany robbery can be found in the *New York Times* articles "The Robbers at Tiffany's: Not Jewelers" (Lynette Holloway, September 7, 1994) and "Was It Temptation at Tiffany's?" (George James, April 18, 1995). *The Diamond District News* (July 2008) provides some coverage of the "10 West diamond." Statistical information on diamond crimes is available at the Jewelers Security Alliance website (www.jewelerssecurity.org). "Diamond

Deals Were a Fraud, Lawsuit Claims" (Anemona Hartcollis, *New York Times,* June 29, 2007) mentions the charge against a father-and-sons team, regarding a stolen $3.3 million worth of diamonds. The writings of Lubin, Shield, Richman, Schumach, Bates, and others aided my understanding of arbitration. Shield, in particular, discusses the high regard in which the institution is held by the New York State court system. Articles from the *New York Times* such as "4th and Last Suspect in Murder of Jeweler Surrenders to Police," by Glen Fowler (April 15, 1986); "Jeweler Found Shot to Death in His Diamond District Office," by Leo Brodsky (May 5, 1986); "Police Seeking a 4th Suspect in Slaying of Jeweler" (April 14, 1986); "2 Employees Are Killed During Jewelry Holdup," by Elizabeth Kolbert (September 27, 1985); and "$1.5 Million in Gold and Gems Stolen from Diamond Dealer" (April 16, 1985) supplemented my research on the state of crime in the diamond district during the eighties. "Diamond District Thieves Use Sledgehammers to Smash Windows, Grab Loot Daylight Dare," by Kerry Burke (*Daily News,* October 3, 2007), is a source for the daytime burglary of an exchange on Forty-seventh Street through its windows. The term "finger job" comes from Lubin's history of the Diamond Dealers Club.

Chapter 12: Blood in the Land of Diamonds

Shield, Hart, "The Nature of Diamonds Exhibition Notes," Dickinson, and Campbell write about diamonds' Indian past. *The Nature of Diamonds* contains translations and discussions of portions of the legend of Bala from the fifth-century document *The Ratnapariksa.* "The Nature of Diamonds Exhibition Notes," Dickinson, Hart, Shield, Cockburn, Campbell, and Rushby (who authored *Chasing the Mountain of Light* [St. Martin's Press, 2000]) write on the subjects of diamond mythology, lore, superstition, and early beliefs. Another useful article on this topic is "From Medieval Lords to Modern Lovers," by Gavanndra Hodge (*Independent on Sunday,* March 24, 2002). My study of blood diamonds, artisanal mining, and Fair Trade diamonds was formed and informed by various people, books, and articles. For further reading, consult www.diamonds.net, Greg Campbell's book (which helped educate me on some of the figures pertaining to smuggling), Tom Zoellner's book, Matthew Hart's book, Douglas Farrah's writings (especially on the connection between al Qaeda and blood diamonds), and Andrew Cockburn's article "Diamonds: The Real Story." Campbell, Hart, and Zoellner write extensively about the African wars that produced blood diamonds. The Fatal Transactions campaign is covered in Hart's book. The quote beginning "the increasing outflow of Angolan diamonds" appears on page 8 of *A Rough Trade.* (Global Witness, 1998). The Nelson Mandela quote beginning "If there is a boycott of diamonds" comes from "Mandela Concerned at Diamond Boycott" (www.irinnews.org, the UN Office for the Coordination of Humanitarian Affairs). Hart notes that the DTC's buying offices in Antwerp ceased their purchasing of rough diamonds on the open market by 2000. In addition to my conversations with Martin Rapaport, I learned about his involvement in the fight against conflict diamonds from some of his writings, including "Ethical Nightmares" (*British Vanity Fair,* June 2006). The quote beginning "to handle PR" appears on page 130 of Campbell's book.

World Vision provided me with a DVD of the Martin Sheen television spot. The Kimberley Process website (www.kimberleyprocess.com) contains useful information on the history and particulars of the monitoring system. The influential "Diamonds in the Rough" (Human Rights Watch, 2009) reports on the events that took place at the Marange diamond fields. Many news articles from around the world, such as "Mugabe Faces Blacklist Over Rogue Diamonds," by Jan Raath (*The Times,* London, June 11, 2007), have tracked the situation in Zimbabwe and the reaction of the diamond industry and NGOs. The *New York Times* article I refer to in this chapter, which publicized Rapaport's warning to members intentionally trading Marange diamonds, is "Zimbabwe: Some Diamonds Banned," by Celia Dugger (August 17, 2010). A Rapaport press release ("Rapaport Bans Zimbabwe's Marange Diamonds," www.diamonds.net) quotes Martin Rapaport as saying, "The Kimberley Process is being used as a fig leaf to cover up human rights abuses in the diamond sector. . . ." "Human Development Indices: A Statistical Update 2008" (Ed. Green Ink [United Nations Development Programme, 2008]) demonstrates Sierra Leone's place on the Human Development Index. The 2008 Rapaport Fair Trade Conference in Las Vegas and the materials featured on Rapaport's Fair Trade website, including white papers from the Rapaport Fair Trade Conference at JCK 2008 and the Special Edition of the *Rapaport Diamond Report,* taught me a great deal about Fair Trade, its goals, and its challenges. Information about the Diamond Development Initiative is available on their website, www.ddiglobal.org. Ian Smillie, Greg Valerio, and Martin Rapaport helped me understand the philosophy behind Fair Trade and its possibilities. Although his article and book were written in a different context, Ed Epstein's analysis of the modern diamond in "Have You Ever Tried to Sell a Diamond?" also informed my understanding of Fair Trade's potential. To investigate the market's ethical jewelry offerings, I consulted various trade websites such as www.tenthousandvillages.com and www.pristineplanet.com.

Chapter 13: The Shows

Representatives of the AGTA were helpful in supplying details about their show. "The Israeli Who Cracked the Diamond Cartel" (Zev Chafets, *International Herald Tribune,* September 15, 2007) and Andrew Cockburn's "Diamonds: The Real Story" both make the point that Lev Leviev has posed a challenge to De Beers.

Chapter 14: Vegas

Rapaport's price lists contain exact figures to support the fluctuation of prices during 2008. The *JCK Las Vegas 2008 Show Guide* and representatives of the JCK educated me on the particulars of their show. "The Rapaport Breakfast" DVD with which the Rapaport staff supplied me was helpful in recounting the details of the gathering and Martin Rapaport's State of the Industry address.

Chapter 15: The Diamond Growers

Drs. Hemley and Yan educated me on the science of diamonds, and taught me just about everything I know about diamond growing. In addition to my visit to the Geophysical Laboratory, I was introduced to the subject of man-made diamonds through Ulrich Boser's thorough article "Diamonds on Demand" (*Smithsonian Magazine*, June 2008) and "The Nature of Diamonds Exhibition Notes" website. Boser writes about potential and current uses of synthetic diamonds. Robert M. Hazen's *The Diamond Makers* (Cambridge University Press, 1999) and Kanfer's *The Last Empire* were my main sources on the race for synthetics. Element Six's website (www.e6.com/en/aboutelementsix/boardofdirectors/) notes the presence of two De Beers shareholder representatives on its board of directors. The misconceptions of fourteenth-century alchemists regarding the origin of diamonds were brought to my attention by Greg Campbell's book. Moissanite's website (www.moissanite.com) is the definitive source for all things Moissanite. In "Labs Dull Diamonds' Luster," by Danielle Rosingh, of Bloomberg News (*Financial Post Canada*, June 15, 2007) writes that, according to Apollo, diamond growers create about one billion carats of diamonds each year. Details on the Federal Trade Commission's denial of the petition vis-à-vis the use of the word "cultured" for man-made diamonds are available on the FTC website (www.ftc.gov). The De Beers representative quote beginning "Our research shows that 94 percent of women" appears in the article "Diamonds Lose Their Sparkle," by Tom Whitehead (*The Express*, February 9, 2005). The quote beginning "If you meet a woman whom you are going to spend the rest of your life with" comes from "A Fight to the Glitter End: Diamond Miners Versus Man-Made Gems" (*Weekend Australian*, June 16, 2007). The Martin Rapaport article referenced in this chapter, which contains the quote beginning "We are in a very sophisticated long-term technology race," is entitled "Cultured Diamonds" and appears on www.diamonds.net (October 8, 2003). The story of John Gurney's discovery is related in Matthew Hart's book. Richard Gibson Hubler's article "Diamonds Come to America" elaborates on the involvement of diamonds during World War II. Kanfer writes about De Beers' acquisition of ASEA and its deal with GE. The *Harper's Magazine* quotes referenced in this chapter come from page 467 of the September 1859 issue of the magazine.

Chapter 16: LifeGem

Dean VandenBiesen and the LifeGem website (www.lifegem.com) were my chief guides on the company.

Chapter 17: Safe Crumbs

My translation for the saying "May the Almighty comfort you amongst the mourners of Zion and Jerusalem" comes from Chabad's website (www.chabad.org).

Chapter 18: I Want You to Take Care of Your Daughter

Gary Schuler was a valuable source on auction houses and the transformation of the secondhand industry. Sotheby's auction results are available on their website, www.sothebys.com. For this chapter, I also consulted their catalog, *Important*

Jewels: New York 25 September 2008 (Sotheby's Inc., 2008). For further reading on Lesotho's noteworthy diamond output and the kingdom in general, one might turn to "Lesotho Mine Yields One of the World's Largest Diamonds" (Luke Baker, Reuters, September 21, 2008), "Gem Diamonds' Lesotho Mine Has Real Sparkle" (*Daily Telegraph,* London, January 7, 2009), "Rock Record" (*Independent,* London, December 3, 2008), and "Lesotho: Cultivating the Upside in the Mountain Kingdom" (*Business Day, Africa News,* July 21, 2008). Christopher Mason's book *The Art of the Steal* (Berkley Books, 2005) contains a wealth of information on Alfred Taubman, the history of Sotheby's and Christie's, and the price scandal. The quote "People don't want to come in our door, because we make them feel stupid" appears on page 38 of *The Art of the Steal.* Other articles on the topic include "Return of the Sotheby's Jailbird," by Louise Armistead (*Sunday Times,* London, April 1, 2007), and "Who's Who in the Sotheby's Price-Fixing Trial," by Anna Rohleder (*Forbes,* Connoisseur's Guide). The International Diamond Exchange (www.idexonline.com) provides helpful statistical information on the decrease of jewelry retailers during the recession, and other similar figures.

Chapter 19: The New Dealers

"Diamond District Tenants Worry about Spate of Building Deals," by Terry Pristin (*New York Times,* July 9, 2008), provided data for this chapter. It mentions the exodus of twelve tenants from a Forty-seventh Street building, and the fact that building owners planned to take in non-diamond-dealing businesses. "Diamonds Are Forever—But Districts . . . ," by Lenore Skenazy (*Daily News,* November 19, 2006), and "An Unshowy Setting for Gems," by Christopher Gray (*New York Times,* August 31, 2008), report on the Forty-seventh Street neighborhood since 2006. And several articles, such as "Key Dealer Shuns Diamond Tower Project," by Sophia Chabbot (*WWD,* January 9, 2007), "Metro Briefing New York: Manhattan: Diamond Club to Stay Put," by Charles V. Bagli (*New York Times,* January 9, 2007), and "A Developer's Vision for the Diamond District Sends Uneasy Shivers Down 47th Street," by Charles V. Bagli (*New York Times,* March 13, 2006), cover Barnett's International Gem Tower. "A Developer's Vision" refers to Mr. Barnett's other name, Gershon. "Diamond District Tenants Worry" makes note of Barnett's motives regarding his involvement in the diamond district. The GIA's plans to transfer to the International Gem Tower is announced in a press release ("NY GIA Lab to Relocate to International Gem Tower," GIA, www.diamonds.net, January 14, 2010). The Indian Diamond and Colorstone Association website (www.idcany.com) provides information on Indian diamond and jewelry businesses in America.

Selected Bibliography

Bates, Rob. "DDC Celebrates Diamond Jubilee: An Oral History." *Jewelers Circular Keystone* 177, no. 7 (July 2006): 100–102, 104, 106, 108, 110.

Beah, Ishmael. *A Long Way Gone: Memoirs of a Boy Soldier.* New York: Sarah Crichton Books, 2007.

Bernstein, Lisa. "Opting Out of the Legal System: Extralegal Contractual Relations in the Diamond Industry." *Journal of Legal Studies* 21, no. 1 (January 1992): 115–57.

Boser, Ulrich. "Diamonds on Demand." *Smithsonian Magazine,* June 2008, p. 52.

"Botswana Country Profile." *BBC News,* July 27, 2010. http://news.bbc.co.uk/.

Brazilian, Alexa. "Hope Diamonds." *Elle,* December 2008, 106.

Campbell, Greg. *Blood Diamonds: Tracing the Deadly Path of the World's Most Precious Stones.* Boulder, Colo.: Westview Press, 2004.

Clark, Donald S., secretary, Federal Trade Commission, to Cecelia L. Gardner, president and CEO, Jewelers Vigilance Committee, et al., July 21, 2008. www.ftc.gov/os/2008/07/G711001jewelryguides.pdf.

Clean Diamond Trade Act, H.R. 1584, 108th Cong. (January 2003).

"Clean Diamond Trade Act," Public Law no. 108-19, 25 April, 2003, 117 Stat 631 (2003).

Cockburn, Andrew. "Diamonds: The Real Story." *National Geographic* 201, no. 3 (March 2002).

"Color Grading D-to-Z Diamonds at GIA." Adapted by Russell Schor. Gemological Institute of America (GIA), May 2009. Special Condensed Article from the Winter 2008 issue of *Gems & Gemology.*

"Conflict Diamonds Sanctions and War." Published by the UN Department of Public Information in cooperation with the Sanctions Branch, Security Council Affairs Division, Department of Political Affairs. Updated March 21, 2001. www.un.org/peace/africa/Diamond.html.

Congressional Record-Senate. S2011-S2015 (March 18, 2002).

"Cote D'Ivoire: Conflict Diamonds Continue." *Other Facets,* Partnership Africa Canada, Number 27, June 2008. http://www.pacweb.org/Documents/Other-Facets/OF27-eng.pdf.

Cuellar, Fred. *How to Buy a Diamond.* 3rd, 5th, 6th editions. Naperville, Ill.: Sourcebooks Casablanca, 2008.

Diamond District News 7, no. 9 (September 2008); 7, no. 10 (October 2008); 7, no. 11 (November 2008); 7, no. 12 (December 2008); 8, no. 3 (March 2009); 8, no. 4 (April 2009). Ed., Nicole Caldwell. New York: DDN Publishing, Inc.

The Diamond Road. Produced by Kensington Communications, directed by Nisha Pahuja and Manfred Becker, written by John Kramer and Nisha Pahuja, released July 29, 2007.

"Diamonds in the Rough: Human Rights Abuses in the Marange Diamond Fields of Zimbabwe." *Human Rights Watch,* June 26, 2009.

Dickinson, Joan Y. *The Book of Diamonds.* New York: Dover Publications Inc., 2001.

Epstein, Edward Jay. "Have You Ever Tried to Sell a Diamond?" *The Atlantic Monthly,* February 1982.

———. *The Diamond Invention.* www.edwardjayepstein.com/diamond/prologue.htm. London: Hutchinson, 1982. (Originally published by Simon & Schuster in 1982 as *The Rise and Fall of Diamonds.*)

———. *The Death of the Diamond.* London: Sphere Books Limited, 1983.

Farah, Douglas. "The Role of Conflict Diamonds and Failed States in the Terrorist Financial Structure." October 24, 2003. http://www.douglasfarah.com/articles/conflict-diamonds.php.

———. *Blood from Stones: The Secret Financial Network of Terror.* New York: Broadway Books, 2004.

Friedman, Jane. "Women: New Facet of the Diamond Trade." *New York Times,* February 18, 1979, p. F2.

"GIA Master Stones: Only Precision Will Do." Adapted by Russell Shor. Gemological Institute of America (GIA), May 2009. Special condensed article from the Winter 2008 issue of *Gems & Gemology.*

"Guides for the Jewelry, Precious Metals and Pewter Industries." Federal Trade Commission. 16 CFR Part 23. *Federal Register,* Vol. 64, no. 109. June 8, 1999. Proposed Rules. Pp. 30448–50.

Hall, Tony, with Tom Price. *Changing the Face of Hunger.* Nashville: Thomas Nelson, 2006.

Harlow, George E., ed. *The Nature of Diamonds.* Cambridge: Cambridge University Press, 1998.

Hart, Matthew. "How to Steal a Diamond." Foreign Affairs, *The Atlantic Monthly*, March 1999.

———. *Diamond: The History of a Cold-Blooded Love Affair.* London: Fourth Estate, 2003.

Hazen, Robert M. *The Diamond Makers.* Cambridge: Cambridge University Press, 1999.

"Human Development Index, The (HDI)." UN Development Programme (UNDP) website. http://hdr.undp.org/en/statistics/hdi.

Kanfer, Stefan. *The Last Empire: De Beers, Diamonds, and the World.* New York: Farrar, Straus & Giroux, 1995.

Kolish, Elaine D., of Sonnenschein Nath and Rosenthal LLP, to Donald S. Clark, secretary, Federal Trade Commission, December 11, 2006. www.jvclegal .org/SFX4651.pdf.

"Loupe Holes: Illicit Diamonds in the Kimberley Process." Global Witness and Partnership Africa Canada, November 2008. www.pacweb.org.

Lubin, Albert. *Diamond Dealers Club: A Fifty-Year History.* New York: Diamond Dealers Club, Inc., 1982.

"Martin Sheen: Sierra Leone—Conflict Diamonds PSA." Conflict diamonds TV spot, produced by World Vision and GMMB.

Mason, Christopher. *The Art of the Steal: Inside the Sotheby's-Christie's Auction House Scandal.* New York: Berkley Books, 2005.

Meredith, Martin. *Diamonds, Gold, and War: The British, the Boers, and the Making of South Africa.* New York: Public Affairs, 2007.

Michman, Dan. "Study Sheet on the Weekly Torah Portion; *Daf Shvui No. 172*; Parashat Tezaveh 5757." Bar-Ilan University. From "*Turquoise, Sapphire and Diamond" to "Mazal U'brachah": Jews in the Diamond Trade in Modern History,* by Dieter Pschichholz, FGA. Diamant. Stuttgart: Rühle-Diebener-Verlag, 1979.

N. W. Ayer Advertising Records, Collection 59, Box 40, Folders 4, 5, 6, 7, and 8. Archives Center, National Museum of American History.

Partnership Africa Canada. *Other Facets,* no. 33 (August 2010). www.pacweb .org/index-e.php.

Rapaport Diamond Report Price List 31, no. 19 (May 16, 2008); 31, no. 20 (May 23, 2008); 31, no. 36 (September 12, 2008); 31, no. 37 (September 19, 2008); 31, no. 38 (September 26, 2008); 31, no. 42 (October 31, 2008). Approximate High Cash Asking Price Indications.

Rapaport Diamond Report 28, no. 30 (August 5, 2005).

Rapaport, Martin. "Ethical Nightmares." *Rapaport Diamond Report Special Edition* 30, no. 1 (January 5, 2007): 7.

———. "Spiritual Sparkle." *Rapaport Diamond Report Special Edition* 29 (June 1, 2006). (Also appeared in the June 2006 *British Vanity Fair* special supplement under the title "Rough Justice.")

Resolution 1173 (1998). UN Security Council, Distr. General, S/Res/1173 (1998). Adopted by the Security Council at its 3891st meeting on June 12, 1998.

Resolution 1176 (1998). UN Security Council, Distr. General, S/Res/1176 (1998). Adopted by the Security Council at its 3894th meeting on June 24, 1998.

Richman, Barak D. "Community Enforcement of Informal Contracts: Jewish Diamond Merchants in New York." Harvard John M. Olin Discussion Paper Series. Paper No. 384, 9/2002. www.law.harvard.edu/programs/olin_center/.

———. "How Community Institutions Create Economic Advantage: Jewish Diamond Merchants in New York." *Law and Social Inquiry: Journal of the American Bar Foundation* 31, no. 2 (Spring 2006): 383–420. Blackwell Publishing, 2006, American Bar Foundation. (Note: an earlier version of this article was also consulted.)

———. "Ethnic Networks, Extralegal Certainty, and Globalization: Peering into the Diamond Industry." Duke Law School Legal Studies Research Paper Series, Research Paper No. 134, December 2006. Forthcoming in *Legal Certainty Beyond the State,* ed. Volkmar Gessner.

Roberts, Janine. *Glitter & Greed.* Revised ed. New York: The Disinformation Company Ltd., 2007.

"A Rough Trade: The Role of Companies and Governments in the Angolan Conflict," January 12, 1998. London: Global Witness. www.globalwitness.org/media_library_detail.php/90/en/a_rough_trade.

Rushby, Kevin. *Chasing the Mountain of Light.* New York: St. Martin's Press, 2000.

Salmans, Sandra. "A Diamond Maverick's War with the Club on 47th Street." Financial Desk, *New York Times,* November 13, 1984, p. A1.

Schor, Russel. *Connections: A Profile of Diamond People and Their History.* Israel: Published for The World Federation of Diamond Bourses by International Diamond Publications, Ltd., 1993.

Schumach, Murray. "The City's Most Exclusive Club." *New York Times,* May 6, 1979, p. 255.

———. *The Diamond People.* New York: W. W. Norton & Company, 1981.

Schuster, Karolyn. "Talking Conflict Free with Customers." *Rapaport Diamond Report Special Edition* 30, no. 1 (January 2007): 7.

"Sharing an Article from Martin Rapaport: 'Guilt Trip'—Hon. Tony P. Hall (Extensions of Remarks—May 19, 2000)." *Congressional Record,* House of Rep. 106th Cong. (May 19, 2000). Hon. Tony P. Hall of Ohio.

Shield, Renée Rose. *Diamond Stories: Enduring Change on 47th Street.* Ithaca, N.Y.: Cornell University Press, 2002.

Smillie, Ian, Lansana Gberie, and Ralph Hazleton. "The Heart of the Matter: Sierra Leone, Diamonds & Human Security." Ottawa: Parternship Africa Canada, 2000.

Utting, Francis Arthur James. *The Story of Sierra Leone*. Fla.: New World Book Manufacturing Co., 1971.

Weber, Lauren. "The Diamond Game, Shedding Its Mystery." Money and Business/Financial Desk, *New York Times*, April 8, 2001, p. C1.

"Zimbabwe: End Repression in Marange Diamond Fields." Human Rights Watch News. June 26, 2009.

Websites

Blood Diamond movie website. blooddiamondmovie.warnerbros.com.

The De Beers Group website. www.debeersgroup.com: "Forevermark Launches Consumer Website www.forevermark.com"; "Supplier of Choice"; "The DTC Announces Sightholder List"; "Lifecycle Planning"; "WWF-Lonmin Award"; "Mining Methods"; "Alluvial Mining"; "Marine Mining"; "Early Stage Exploration"; "Advanced Exploration"; "Group Exploration Activities"; "Angola"; "Botswana"; "Central Kalahari Game Reserve (CKGR)"; "Democratic Republic of Congo"; "Open-pit Mining"; "Underground Mining"; "Mining Operations"; "Operating and Financial Review, 2008"; "Damtshaa"; "De Beers Marine Namibia"; "Elizabeth Bay"; "Finsch"; "Jwaneng"; "Kimberley"; "Letlhakane"; "Mineral Processing"; "Comminution and Liberation"; "Concentration"; "Recovery"; "Fine and Coarse Residue Disposal"; "Diamond Value Management"; "Sales and Distribution"; "Diamond Trading Company (DTC)"; "Diamond Trading Company Botswana (DTCB)"; "Namibia Diamond Trading Company (NDTC)"; "Inside De Beers"; "Overview and Principles"; "Sightholders"; "Sorting and Valuing"; "Purchasing"; "Sorting Process"; "Sorting Technologies"; "Sights and Sightholders"; "Best Practice Principles"; "Beneficiation"; "Alrosa Purchasing Commitments"; "Governance"; "Board of Directors"; "The Family of Companies"; "De Beers Canada"; "De Beers Consolidated Mines"; "De Beers Diamond Jewellers"; "Debswana"; "Diamdel"; "Diamond Trading Company South Africa (DTCSA)"; "Element Six"; "Forevermark"; "Namdeb"; "Integrity Is Everything"; "Approach"; "Economics"; "Contribution to Economies"; "Investment in People and Infrastructure"; "Diversification"; "Planning for Closure"; "Ethics"; "Business Principles"; "Communities"; "The De Beers Group"; "Environment"; "Energy and Climate"; "Water Management"; "Materials and Waste"; "The De Beers Fund"; "Principles and Policies"; "De Beers and Petra Announce Purchase of Cullinan Diamond Mine."

Diamond Dealers Club website. www.nyddc.com.

Diamond Facts website, by the World Diamond Council. www.diamondfacts .org.

Element Six website. www.e6.com.

Forevermark website. www.forevermark.com/en/.

The 47th Street BID website. www.diamonddistrict.org/.

Gemesis website. www.gemesis.com.

Gemological Institute of America (GIA) website. www.gia.edu.

Human Rights Watch website. www.hrw.org.

Jewelers Alert website. www.jewelersalert.com.

Jewelers Circular Key Stone website. www.jckonline.com.

Jewelers' Security Alliance website. www.jewelerssecurity.org.

Kimberley Process website. www.kimberleyprocess.com.

LifeGem website. www.lifegem.com.

Moissanite website. www.moissanite.com.

"The Nature of Diamonds." American Museum of Natural History website. www.amnh.org/exhibitions/diamonds/index.html.

Rapaport Fair Trade website. www.diamonds.net/fairtrade/.

Rapaport News on www.diamonds.net: Blauer, Ettagale, "Big-Money Blues," June 1, 2008; Ehrenwald, Jerry, "No Misconception—Lab Created Diamonds Are Distinguishable," July 17, 2008; "Essential Elements of an International Scheme of Certification for Rough Diamonds with a View to Breaking the Link Between Armed Conflict and the Trade in Rough Diamonds," March 20, 2002; Extell, "International Gem Tower Reps to Visit Ramat Gan" (News Release), April 9, 2010; GIA (Gemological Institute of America), "NY GIA Lab to Relocate to International Gem Tower; Institute Purchases Full Floor in Extell's New Industry Tower on 47th St.," January 14, 2010; Extell Development Company (Press Release), "Top Israeli Diamond Firms Purchase Space in IGT," January 17, 2010; Kazi, Zainab S., and Avi Krawitz, "GJEPC Criticizes DTC for Fewer Sightholders," January 2, 2008; Krawitz, Avi, "Graff Buys 478ct Letseng Gem for $18M," December 2, 2008; ———, "KP-Certified Marange Diamonds Auctioned in Harare," August 12, 2010; Michelle, Amber, "Observations, August 2009," August 1, 2009; ———, "Observations, July 2008," July 1, 2008; Miller, Jeff, "FLO, ARM Certify Fairtrade and Fairmined Gold Standards," April 30, 2010; ———, "Global Witness Praises Zimbabwe Suspension Measure," July 31, 2009; ———, "JVC: Use of Term 'Cultured' Must Be Clarified," July 22, 2008; ———, "Suspension of Diamond Exports Wouldn't Impact Zim's Recovery Says NGO," August 3, 2009; ———, "Zimbabwe Spared by KP," November 5, 2009; "Quotes," Rapaport Diamond Conference 2007; Rapaport, Martin, "Boom and Bust," June 1, 2008; ———, "Bottoms Up," March 9, 1999; ———, "Boyajin's Vision Lives," January 5, 2007; ———, "Cultured Diamonds," October 8, 2003; ———, "A Diamond Is a Diamond?" October 8, 2003; ———, "Guilt Trip," April 7, 2000; ———, "Martin Rapaport Statement on Blood Diamond Movie," December 10, 2006; ———, "Speculation," March 1, 2008;

———, "Trust," May 6, 2002; ———, "Tsunami Warning," July 1, 2008; ———, "Understanding the Kimberley Process," January 5, 2007; Serdyukova, Anastasia, "KP Reaches Agreement with Zimbabwe," August 1, 2010; Smillie, Ian, "Conflict Diamonds: What Percentage?" Partnership Africa Canada, January 2007; "Trade Alert: Marange Diamonds," August 12, 2010; "Trade Alert: Rapaport Bans Zimbabwe's Marange Diamonds," November 19, 2009; "World Diamond Congress Resolution: Kimberley Process—October 29, 2002"; World Diamond Council Press Release, "WDC Supports Immediate Audit of Marange Diamonds," November 9, 2009; Yonick, Deborah, "Can KP Stay Viable?" July 1, 2010.

World Diamond Council website. www.worlddiamondcouncil.com.

World Federation of Diamond Bourses website. www.worldfed.com.

Acknowledgments

I am grateful to my agent, Julie Barer, and my editor, Alexis Gargagliano for their aesthetic integrity, enthusiasm, and passion. Thank you, as well, to everyone at Scribner.

Thanks to all of my professors and classmates at the Columbia MFA program. I am indebted to the David Berg Foundation Fellowship.

Martin Rapaport was an infinitely valuable source of knowledge on all aspects of the industry. Thank you to the entire Rapaport team. Thank you to Ian Smillie and Greg Valerio; to David Abraham, Ken Kahn, Andrea Sperling, Morgan Philips, and James Brown. Thank you to Isaac Herschkopf.

Several people supplied me with expert knowledge in various fields directly or indirectly related to the one in which I was immersed. Thank you to Margreta de Grazia, Sean Keilen, and Andrew Campbell.

Thank you to Sue, Jerry, and Rachel for their love and support.

There are too many members of the industry with whom I spoke to name individually, but I am filled with gratitude toward them all.

My parents and grandparents were titanic resources.

Uri: reader, partner, and everything else.

In memory of Steven Oltuski.

About the Author

Alicia Oltuski received her bachelor's and master's degrees from the University of Pennsylvania and an MFA in writing from Columbia University, where she was awarded a David Berg Foundation Fellowship. Her work has appeared on NPR's *Berlin Stories,* and in *The Faster Times, The Bulletin* in Philadelphia, and other publications. She has taught at the University of the Arts and lives in the Washington, D.C., area with her husband. Visit her website at www.aliciaoltuski.com.